Biomass, Bioproducts and Biofuels

Jorge M.T.B. Varejão
Department of Exact Science
Agrarian School/CERNAS/I. I. A.
Polytechnic Institute of Coimbra
Portugal

With contribution from

Chapter 7

**Ferreira, Joana D., Martins, Clara B.,
Assunção, Mariana F.G. and Santos, Lilia M.A.**

Coimbra Collection of Algae (ACOI)
Department of Life Sciences
University of Coimbra, Portugal

CRC Press
Taylor & Francis Group
Boca Raton London New York

CRC Press is an imprint of the
Taylor & Francis Group, an **informa** business

A SCIENCE PUBLISHERS BOOK

First edition published 2022
by CRC Press
6000 Broken Sound Parkway NW, Suite 300, Boca Raton, FL 33487-2742

and by CRC Press
4 Park Square, Milton Park, Abingdon, Oxon OX14 4RN

Library of Congress Cataloging-in-Publication Data (applied for)

ISBN: 978-0-367-35408-4 (hbk)
ISBN: 978-1-032-12424-7 (pbk)
ISBN: 978-0-429-34054-3 (ebk)

DOI: 10.1201/9780429340543

Typeset in Palatino
by Radiant Productions

To Gonçalo and André

Preface

At the beginning of the third decade of the 21st century, Humanity is under pressure from the scarcity of energy resources, compounded by the great environmental damage caused to Earth by fossil materials already burned. The world population approaches 10 billion people, each with the ambition of a better standard of living that is currently being obtained mainly at the expense of using even more fossil materials. This has been happening since the 1970s last century and has led to major climate changes, which are becoming more evident each year and leading to damage and economic loss worldwide. The change in the energy supply model is underway, but there is still sometime before it transforms into a greener version. In the meantime, any new technology that helps to reduce carbon dioxide emissions is welcome as it contributes to lessen the gravity of the situation. There placement of electric cars is a beginning in such an effort and its use is expected to accelerate in the coming years. However, in some cases, electricity is produced from burning fossil fuels, annihilating or aggravating the environmental effort. Greener electricity sources, such as wind, sea waves and all possible renewable energy sources, are required to solve the problem. Other areas will eventually under go a longer transition period due to technical difficulties in replacing fossil fuels, e.g., commercial air travel.

Biomass is a renewable source that can contribute to replacing the use of fossil materials without contributing to net emissions, since its growth occurs through the absorption of CO_2 from the atmosphere and not through the subsoil extraction of carbon-based substances. Biomass has the potential to be converted into liquid fuels, domestic and industrial solvents, plastics, textile fibers and almost all the goods that modern society requires. However, this objective will only be effective if the prices of these new raw material variants are competitive in relation to fossil carbons, which in turn requires a strong effort in research and some period of evolution.

Biomass exists everywhere, but often in places and terrains that make its collection difficult or in accessible by machine tools. Obtaining it with human labour makes the price of any derivative very high and uncompetitive, with exception for small niches of biomass type, used for example in the production of specialized foods. Extensive agricultural activity is one of the main sources of competitive biomass, such as the production of cereals and the cultivation of sugarcane for the production of sucrose, which collects and concentrates large quantities of biomass residues in specific locations. Other sources of biomass that may be relevant to use as raw material are forest and aquatic biomass, such as micro and macroalgae.

A high volume of biomass is already used as a raw material for a wide range of industries, such as cellulosic pulp factories, preparation of heating briquettes, oriented strand board (OSB), medium-density fiberboard (MDF), and particle board (PB), for general use and furniture industry. Most of these uses increase with population growth, which means that the demand for terrestrial biomass will expand incoming years.

The ban on burning grain biomass residues, driven by law in many countries, has led to a strong research effort for the production of liquid fuels from biomass, namely ethanol—which due to its disinfectant properties is a highly desirable end-product, as alcohol is now difficult to find due to the spread of the Covid-19 virus worldwide. Also ethanol can be used as a fuel in explosion motors.

In Europe, the addition of biodiesel to diesel was introduced in the early years of this century as a way of reducing carbon particle emissions of the predominant diesel engines. This policy, instead, has a negative effect on the environment since its use is beneficial from emissions point of view but has led to deforestation to expand oil seed crops production. The net result was greater environmental damage in global terms, which was added to what was already occurring due to the high demand for vegetable oils for the food industry. Restrictions on the preparation of biodiesel with lipid coming from distant sources and the favoring of the conversion of waste frying oils was later tried as a corrective measure.

Meanwhile, the use of biomass-derived bioproducts, such as xylitol, microcrystalline cellulose (MCC), carbon fiber-like materials, sugars, antioxidants, etc., has begin to increase, which means that strategies must be devised to efficiently process all available biomass with a view to its full use. Other new sources of biomass such as micro- and macroalgae have started to gain importance. Microalgae biomass is rich in constituents that can easily be converted into biofuels and bioproducts, making it a complementary source to terrestrial biomass.

This book deals with multiple attempts, developed over decades of work, to use biomass as raw material for obtaining bioproducts and biofuels. Many of the techniques described are new and have not yet been published. They can be developed and converted into useful processes. The main objective of the book is to offer new perspectives on the use of biomass, with a view to contributing to the reduction of fossil carbon dioxide emissions through the development of efficient biorefineries.

A work like this rests on the author's dedication to this field of research for many years, either through the teaching of subjects in the Biotechnology Degree course, namely the Biotechnology and Biorefinery unit, or by the orientation of graduate students' of the master's degree, in the scope of the activities of the institution to which the author belongs—the Escola Superior Agrária (ESAC) of the Polytechnic Institute of Coimbra (IPC).

The contribution of students with their effort to carry out experimental research work for the preparation of their dissertations, has provided, over the years, a valuable source of new research possibilities.

Recognition to the research center "Center for Natural Resources, Environment and Society", CERNAS (FCT), Environment and Society group, from which financial support has always been provided for all activities. A sincere thanks to everyone who contributed over the years to the content of this book, in particular to Professor Lilia Santos and researcher Mariana Assunção, authors of Chapter 7; and to all ACOI-Coimbra researchers, Professor Fernando Simões of ISEC/IPC for his assistance in carrying out mechanical tests on composite samples, to researchers Célia Ferreira and Verónica Oliveira for their invaluable contribution, to the staff of the Chemistry and Biochemistry Laboratory and Solos at ESAC, to Stephanie Delgado, Sofia Benedettini, Yasheley van Es, Beatriz Sanchez, Luis Breda, among many others.

June 2020   Coimbra, Portugal

Contents

Chapter 1

Biomass Structure and Disassembling

1. Introduction

Biomass is a renewable organic material that comes from plants, animals and microrganisms. Its origin is diverse and is spread everywhere—forests, agricultural activity and its residues, animals, microorganisms, aquatic biomass, etc. In the latter case, microalgae, which have only recently begun to be better studied, have proved to be very promising in the production of biomass, given the rapid growth that can be done in reservoirs that occupy a very small area of land.

The biomass accumulated in small areas such as agricultural land, etc., is especially interesting, due to its availability in large quantities, examples occur in vegetable residues originating from extensive agricultural production of cereals and sugarcane. These materials represent a portion of the biomass from which derivative products at competitive prices can be obtained, however, biomass collection activity is expensive, unless it is done by mechanical means. Residues from the production of cereals in the food chain with high consumption, such as wheat and rice, are of particular interest as biomass is already being created by the normal processing of the cereal as residue which has no or very limited use.

Special importance is given to these residual biomasses in this book. They have limited use, in Europe mainly as food and as bedding for animals, and their excess is often burned, contributing to the already serious level of carbon dioxide emissions (Xiao et al. 2001). Some countries have already issued regulations and laws that prohibit or oblige to reduce this burning. Often, the biomass is deposited in landfills or simply left in the soil. In 2018, the total amount of residual agricultural and forest biomass produced in the world was estimated at around 10^5 or 100000 billion metric tons, with the largest producers being the USA and Brazil (Kumara and Behera 2018).

Terrestrial biomass was designed by Nature to give plants mechanical resistance and robustness to deal with adverse environmental conditions, while having characteristics necessary for the life and growth of the plant, which involve different mechanisms, an example will be the transport of substances through the tissues.

Wood is an example of a type of biomass used by man in multiple ways, it has good availability, adequate mechanical properties for construction, along time duration, ease of cutting, etc. Pinewood, for example, without any chemical treatment is capable of sustain the roof of a house for more than a century, being only susceptible to attack by several species of microorganisms in the last few years. Despite this strong structure, pines are often not strong enough to deal with adverse weather conditions, such as storms and hurricanes, an example among many others is shown in Figure 1, with the effect on Pinus Pinaster pines from Hurricane Leslie in Coimbra, Portugal, on the night of October 13, 2018 (10:40 pm).

Figure 1. Pinus pinaster tree broken by the strong winds of Hurricane Leslie, on the night of October 13, 2018, in the "Quintada Portela" area, Coimbra, Portugal.

Structurally, biomass is a lignocellulosic material, synthesized mainly in the primary cell wall of plants. The involvement of the secondary wall seems to be at the level of intussusception as the main contribution, since the growth of the plant occurs essentially in the primary wall. Its exact structure is difficult to precisely define. The cell wall has a composite nature consisting of the combination of different materials with different mechanical and physical-chemical properties that are sometimes complementary, making it a particularly resistant material. Biomass has a high polymorphism, depending on the type of plant and, on the plant, its location.

The composite nature of lignocellulosic materials has been known for sometime and its main components are, in decreasing order of mass; cellulose, hemicellulose and lignin. Some types of biomass may have other important secondary constituents, such a spectin, lipids, terpenes and sugars. Table 1 shows the chemical composition of some common types of biomass, in different types of wood, residues from agricultural activity and from different grasses. The relative content of each of the three main constituents is presented.

Both hemicellulose and cellulose are carbohydrates polymers in their nature, when considering their total percentage, values as high as 70–80% of the dry matter can be found in the biomass samples (see Table 1), making it

Table 1. Main component compositions of some examples of lignocellulosic biomass types (adapted from Isikgor and Becer 2015).

Lignocellulosic biomass		Cellulose (%)	Hemicellulose (%)	Lignin (%)
Hardwood	Oak	40.4	25.9	24.1
	Eucalyptus	54.1	18.4	21.5
Softwood	Pine	42.0–50.0	24.0–27.0	20.0
	Scots Pine	40.0	28.5	27.7
Agricultural waste	Wheat Straw	35.0–39.0	23.0–30.0	12.0–16.0
	Barley Hull	34.0	36.0	13.8–19.0
	Barley Straw	36.0–43.0	24.0–33.0	6.3–9.8
	Rice Straw	29.2–34.7	23.0–25.9	17.0–19.0
	Rice Husks	28.7–35.6	12.0–29.3	15.4–20.0
	Oat Straw	31.0–35.0	20.0–26.0	10.0–15.0
	Ray Straw	36.2–47.0	19.0–24.5	9.9–24.0
	Corn Cobs	33.7–41.2	31.9–36.0	6.1–15.9
	Corn Stalks	35.0–39.6	16.8–35.0	7.0–18.4
	Sugarcane Bagasse	42.0	25.0	20.0
	Sorghum Straw	32.0–35.0	24.0–27.0	15.0–21.0
Grasses	Ryegrasses	29.0	30.0	6.7
	Switchgrass	35.0–40.0	25.0–30.0	15.0–20.0

possible infer at first sight its saccharification and subsequent fermentation to produce substances such as ethanol, lactic acid or any other product that can be obtained from sugars. As will be seen below, obtaining sugars, especially from the cellulose component, is far from easy and constitutes the main obstacle to the successful conversion of biomass into a variety of useful products.

The spatial arrangement of the three main components of biomass has been the focus of intense debate in recent decades (Wang and Hong 2016). The problem is relatively complex, considering the many different types of existing biomass, with Nature organizing the constituents in a wide variety of ways to meet the specific needs of plants. These biomass heterogeneity properties present great difficulty in the analytical characterization of the biomass structure, since the analytical techniques available have difficulty in the precise determination of the different components, and their organization is often deduced by indirect measurements.

A large number of simplified models for the structure of biomass have been published in the literature (Li et al. 2016). To add understanding to this problem, a detailed structure of each of the three main components of biomass is discussed and based on this knowledge an update of the biomass structure is proposed.

2. Biomass main components

2.1 Cellulose

A detailed understanding of the cellulose structure is essential for the project that pursues its disassembly into simple sugars or other bioproducts such as a cellulose microcrystalline or nanocellulose. The glycosidic bond prevalent in cellulose is the β(1,4), which causes the next glucose unit to rotate 180° in relation to the previous one. The glucose polysaccharides polymerized through this link form a single homogeneous chain that extends rectilinearly to form a single fibril. On the other hand, the use of the α(1,4) bond in polyglycosides gives rise to a helix structure, for example that found in amylose, a component of starch.

Multiple simple cellulose fibrils are synthesized on the outside of the plant's cell wall in a matrix of several enzymes per cell, the cellulose synthase enzymes (CESA), often called rosettes (Li et al. 2017). The number of individual single fibril which are synthesized simultaneously, range from values as low as 6 to a greater number, as 36 or more. These are organized in microfibrils showing crystalline parts and amorphous zones, the factors that lead to this differentiation are not well known, possibilities are the need to cause curvature in some polymer points or manipulations in CESA rosettes.

The coupling of multiple fibrils gives rise to the formation of macrofibrils that are stabilized through a network of hydrogen bonds between the fibrils. In macrofibrils growth intervention of multiple cells is expected in their elongation and thickening process. The way this is achieved should involve coupling of new sets of microfibrils to those previously made, and is supported by visible details in SEM biomass images where finger-shaped ends are common, suggesting the completion (or beginning) of microfibril assembly at certain points.

The number of individual microfibrils in the macrofibrils can reach thousands. The average length of each individual fiber can be determined by the degree of polymerization (DP), values from 900–6000 (Hallac and Ragauskas 2011) to 15,000 (Mittal et al. 2011) are considered.

Both microfibrils and macrofibrils show crystalline and amorphous zones. The location and direction of production of the polymeric fibrils by the rosette are controlled by the cell, of observable great diversity, depending on the type of plant, cell or organism (algae), with some differences in layout (Huang et al. 2020). Cereals and plants with apical development tend to organize macrofibrils in parallel super-arrays, while wood and other types of cell walls present an overlapping crossed organization of macrofibrils, see below where a model of such an arrangement is presented.

This arrangement gives strength and robustness in several directions to meet the needs of the plant. Figure 2 shows a SEM image of a sample of dried and ground wheat straw (1 mm), partially hydrolyzed with sulfuric acid (60°C), under conditions in which hemicellulose, lignin and amorphous cellulose were hydrolyzed and removed by washing with water.

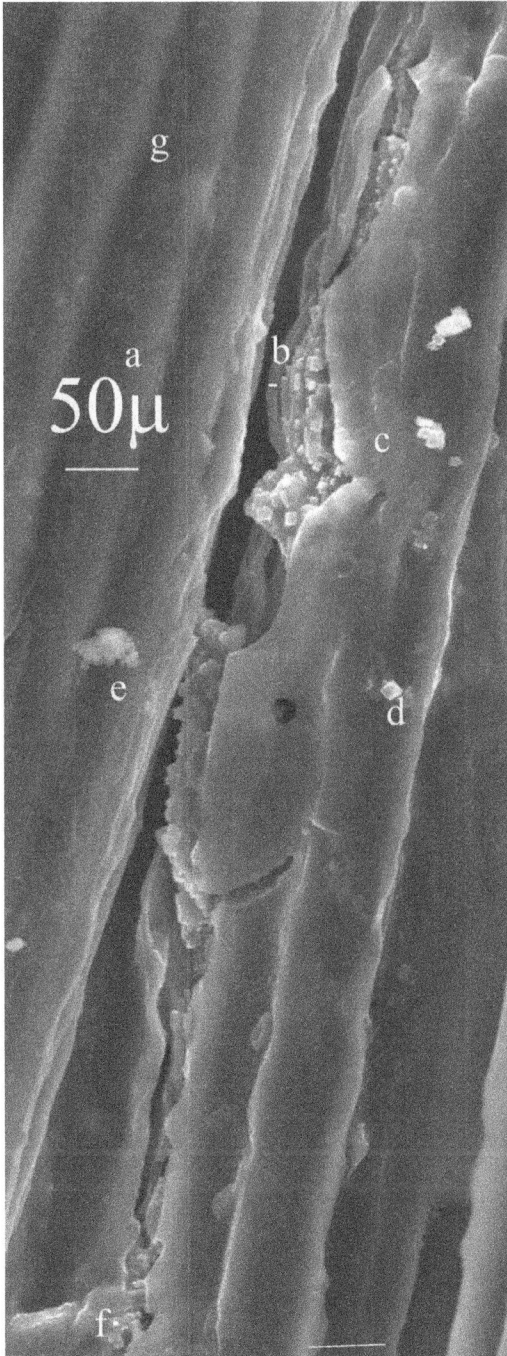

Figure 2. A SEM image of a super-array of wheat straw macrofibrils with some macrofibrils structurally broken by the effect of partial acid hydrolysis, showing their inside microfibrils destroyed. These appear as a set of multiple orthorhombic and monoclinic cellulose crystals.

The image shows relevant details. The macrofibrils themselves (a in Figure 2) are arranged in a super-array where a larger number are positioned in a parallel form, showing diameters ranging from 300 to 500–600 μm. On the left, an intact arrangement of macrofibrils can be seen, with a single macrofibril diameter of about 50 μm (a, in Figure 2). At the right we can see at least one, or more likely several macrofibrils partially destroyed, with their internal content visible. Inside, an arrangement of multiple crystals with different sizes is easily seen, but all of them have a cubic and rectangular orthorhombic appearance (the zones b and f, in Figure 2). Some appear to have a monoclinic geometry and these geometries may be difficult to discern depending on visualization angle. In the literature, monoclinic geometry is often referred to cellulose crystals (Mittal et al. 2011). In zone b, it is possible to clearly observe that the surface of the macrofibrils appears to form a homogeneous coating or an envelope that is certainly due to the effect of acid hydrolytic treatment. This outer layer has a different behavior against acid in relation to the internal content. The same material appears to cover the intact macrofibril on the left (zone g). A possibility frequently mentioned in literature models points to a layer of hemicellulosic sugars on the surface of cellulosic fibers. This is in line with the observations, since C5 sugar are less stable in acidic conditions and can give rise to products such as aldehydes that can polymerize to produce a polymeric homogeneous material, as seen in the image. The size of this coating is approximately 5–8 μm (dimension b in Figure 2).

Within the macrofibrils with the ruptured outer layer, sulfuric acid was able to extensively hydrolyze part of its content, leaving a numerous formation of cubic and rectangular crystals, along the entire length of the macrofibril; both at the top and a the bottom of the image, corresponding to the destroyed macrofibril. Some single crystals or fused crystal formations are seen on the outside of the surface (c in Figure 2), some at some distance from the macrofibril (examples of these are shown respectively in zones e and d, see Figure 2). A detailed measurement of the dimensions of several of the crystals within the macrofibril was made. The crystal measurements, in μm, resulted in dimensions of approximately 3×12; 4×7; 3×3; 5×4; 10×8; 3×3; 18×10. The predominant shape is rectangular with some examples of square structure. This suggests cellulose allomorph I (see below) as the main unit. It is also possible to see crystals fused in the open macrofibril, some of them intertwined (details at the right of b and at the left of c zones). As far as we know, this is the first time that an SEM image has been able to capture a destroyed biomass macrofibril with its partially hydrolyzed content, showing the partially crystalline nature of the internal microfibrils. The size of the crystals belongs to the microscale and taking the smallest and the highest dimensions found as a transversal dimension, it is possible to do a simple accounting to estimate the quantity of microfibril stored inside the macrofibril of wheat straw. The results provide a few hundred (300–600) of microfibrils within the macrofibril. Another consequence of the arrangement of the crystals observed is that the zones that disappeared by hydrolysis must have a different structure (see below).

Obviously, depending on the needs of the fabric, Nature makes variations in the arrangement of the biomass components. In this case, the sample comes from wheat straw, where apical growth is the rule.

Wheat straw cellulose microfibrils show crystalline zones (the zone of crystals seen in Figure 2) and possibly amorphous regions (zones dissolved and absent in Figure 2). Another possibility is that the parts absent may be a different cellulose allomorph, less stable to hydrolysis. Crystalline cellulose is a particularly strong and compact material, with excellent mechanical properties, such as a Young modulus of about 130 GPa and a tensile strength close to 150 GPa (Wu et al. 2010). Obtaining such mechanical properties suggests the possibility of formation of several networks of hydrogen bonds by glucose at the intra- and intermolecular level, reinforced by van der Walls bonds in the direction of the axial fibril stacking. This results in an important structural reinforcement of the final arrangement. A computerized model of octa(D)-β(1,4) oligosaccharide was used as a model structure to visualize these effects, see Figure 3(b) in next section. It presents the result of the calculation of the minimum energy conformation for all chemical bonds, with the incorporation of the most likely hydrogen bonds. In Figure 3, the side and top 3D views of the optimized structure are shown. The most likely formation of hydrogen bonds is deduced between the proximity of –OH groups to the O atoms, allowing a network to be formed. This bonding network essentially involves hydrogen bonds with –OH groups at the equatorial positions of glucose; however, in less frequent cases, hydroxyl groups positioned in axial positions can be used to make additional connections. Because of this, the center of the glucose chain remains essentially hydrophobic, allowing the formation of van der Walls forces in stacking different layers of β-glucose chains. The addition of hydrogen bonds and van der Walls forces makes the overall structure considerable stronger. A detailed study is presented below in the discussion of cellulose allomorphs.

The intensity of the reinforcement can be deduced from the energies of the added hydrogen bond links. The energy content of a single C-C carbon bond is about 80 kJ/mol, and each hydrogen bond has approximately 20 kJ/mol— if we consider the formation of 3 to 5 hydrogen bonds per unit of glucose— both at the inter- and intramolecular level–a 60 to 100 kJ/mol reinforcement is expected in crystalline cellulose in relation to the amorphous form.

These hydrogen bridge networks can take on different arrangements and can be modified by interaction with different types of substances, especially small molecules capable of easily diffusing through the crystalline network. Bigger multi-atom species may be able to interfere if they possess the ability to share hydrogen bonds with the existing network. Examples of substances capable of disturbing the hydrogen bond network are ammonia in gas or liquid phase, alkaline hydroxides, used for example in the mercerization process of cotton, and some acids such as sulfuric or phosphoric acid. The rupture and reorganization of the hydrogen bonding network gives rise to important changes in the structures and properties of cellulose.

2.1.1 Cellulose allomorphs

Understanding the allomorphs of crystalline cellulose is fundamental in any attempt to use cellulose as a raw material for obtaining sugars. It is generally considered that in cellulose from biomass, three or four different types of allomorphs can be found with structural support information in X-ray diffraction analyzes and Fourier transform infrared spectroscopy (FTIR) data, see Figure 3(a). FTIR information sometimes does not allow differentiation of very similar chemical bonds, and clear conclusions are very difficult to validate. Carbon magnetic resonance spectroscopy (13C-NMR), on the other hand, can provide more accurate information on structural details and is very valuable in the field (Goldberg et al. 2015). The construction of computer models helps to understand the different aspects of allomorphs.

In Figure 3 and Table 2 an updated version of known cellulose allomorphs is presented. Allomorph I is found in native biomass and can be observed in two slightly different forms, allomorph Iα and allomorph Iβ. In the dissolution of cellulose and regeneration on precipitation, as occurs in the cotton mercerization process, allomorph II is obtained irreversibly. The treatment of cellulose allomorphs I and II with ammonia gives rise to a different type, the allomorph III.

The types of cellulose allomorphs differ in the organization of the hydrogen bonding network. The hydroxyl groups at positions 1 and 4 of glucose are involved in the formation of covalent bonds in the polymer chain, leaving hydroxyl groups at positions 2, 3 and 6 free, as those capable of forming hydrogen bonds. To these should be added the oxygens of the acetals present in the structure and responsible for the glycosidic bonds as capable to be involved in hydrogen bond formation. In Figure 3(b) a segment of a computer model of the oligosaccharide (D)-β(1,4)-octaglucose is shown, highlighting the relative position of each oxygen atom in the minimum energy conformation. Each glucose unit is rotated 180 degrees relative to the previous one in the chain, causing the glucose hydroxyl(–OH) at position 6 to result close to the hydroxyl at position 2 of the next glucose unit, in the oligomer chain. The glucose hydroxyl in carbon 3 is close to the acetal oxygen responsible for the cyclic form of glucose, the glucose hydroxyl in the carbon 5. A hydrogen bond forms, however, weaker than that formed between glucose positions 2 and 6, which is identical to that formed in alcohols. In this case, oxygen is replaced by an alkyl group, the inductive effects of which decrease the nucleophilicity of oxygen leading to a slightly weaker hydrogen bond (Yu and Klemperer 2005). After the results of Lommerse (Lommerse et al. 1997), while a hydrogen bond HO-H in alcohol has an energy content of about 24 kJ/mol, using ether as an approximation for one of the acetal oxygen in H-(ORR') the energy of the hydrogen bond should contain energy values close to or less than 21 kJ/mol. This reveals that, when making hydrogen

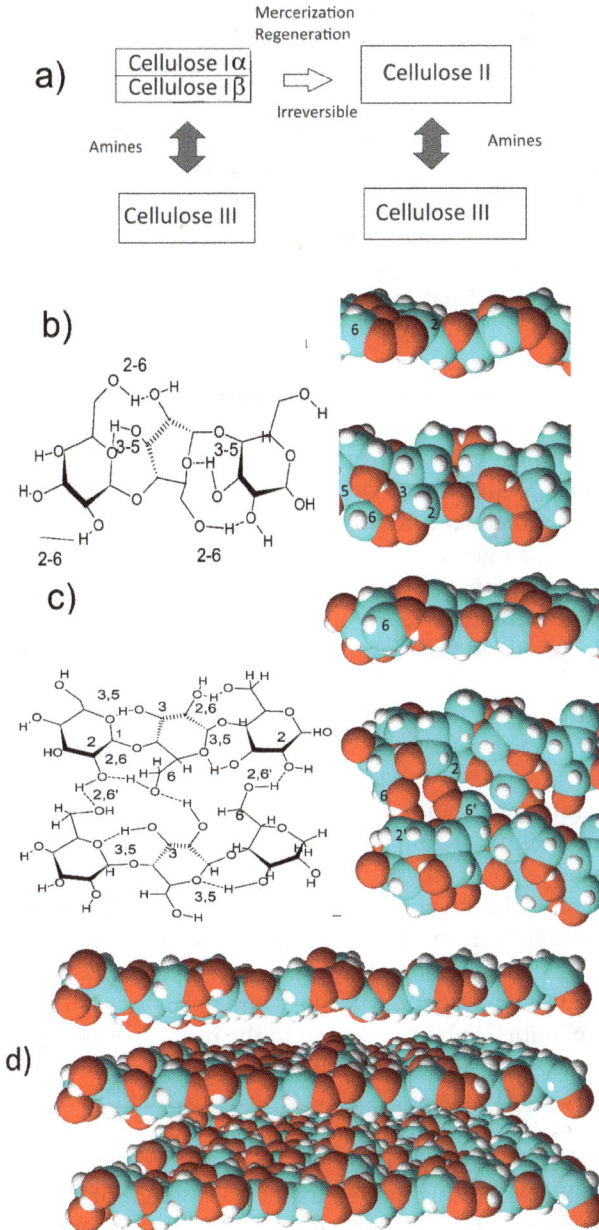

Figure 3. (a) The four accepted cellulose allomorphs and their interconversion processes; regeneration refers to cellulose dissolution followed by precipitation; (b) a segment of octa(D)-β (1,4) glucose oligosaccharide, with minimized energy, showing the most likely formation of hydrogen bonds (see text); (c) the same simulation, but this time with two simple fibers aligned in parallel, showing the possibility of intermolecular hydrogen bond formation (see text for details); (d) At least in cellulose allomorphs type I, as hydrogen bonds are formed mainly at equatorial positions, the glucose units keep an essentially hydrophobic environment in the axial directions. This leads to the formation of sheets that form a crystal lattice, stacking the sheets using van der Walls forces.

bonds between two adjacent 180 degrees inverted glucose monomers, the most favored occur at the positions 2-6 and 3-5. These hydrogen bonds are at the intramolecular level and are represented as dotted lines in Figure 3(b), on the left. The hydrogen bonding angles are in the range of 120 to 150° (Ghanghas et al. 2020). A minimized energy structure with these new chemical bonds produces a very flat ribbon-shaped oligomer. The "ribbon" corresponds to a single β(1,4) glucose microfibril. A 3D view of such an optimized structure is seen in Figure 3(b) in the top and side views. The hydroxyl groups of glucose occupy mainly the equatorial positions. The central zone of glucose remains mainly hydrophobic in nature, with the acetal oxygen responsible for the glucose cycle as the only atom positioned in that zone.

The equatorial hydroxyl groups can make hydrogen bonds with other ribbons located in lateral positions, through the glucose hydroxyl groups in the carbon 2 and 6. To clearly identify these hydrogen bonds from the intramolecular ones an apostrophe is included in the glucose carbon position to indicate atom from other chain, e.g., hydrogen bond 2-6'. Thus, these hydroxyl groups are involved simultaneously in making intra-and intermolecular hydrogen bonds resulting in the grouping of a large number of "ribbons" in one plane. The result is a flat sheet of cellulose. As an example of such an arrangement, see Figure 3(c). In these, the hydrogen bonding network and optimized 3D views are shown. When several of these sheets are stacked, with van der Walls forces between the sheets, the cellulose crystal structure begins to appear. If the glucose "ribbons" occupy a position exactly parallel to each other, the structure of the crystal either monoclinic or orthorhombic.

This structure corresponds to the alomorph Iβ of cellulose, which presents the form of monoclinic crystal (Wu et al. 2018) and is found mainly in native plant cellulose (Song et al. 2015). Different analytical techniques suggest that in cellulose I, the crystals can also be found in the Iα (triclinic) form. Allomorph Iα differs from allomorph Iβ by the relative position of simple fibrils in the adjacent position. This arises due to the possibility of different hydrogen bonding networks. The adjacent fibril may slide, leaving each glucose unit of a fibril to be positioned in the middle of two glucose units of the adjacent one, if this arrangement is repeated it results in the formation of a triclinic crystal (Wada et al. 2001). The angle that this slide originates is about 67 degrees, judged by X-ray crystallography data (Kontturi et al. 2015). In cellulose I the stacking of cellulose sheets to form crystals involves only van der Walls forces. In this model, the establishment of hydrogen bonds is restricted to the zone of the equatorial hydroxyl groups between the adjacent fibrils, see Figure 4.

Allomorph II is considered to have a monoclinic crystalline structure with antiparallel chain orientation, with respect to the final reduction sugar end for non-reduction direction and can only be obtained by dissolution followed by subsequent precipitation of cellulose (Kontturi 2015). This is what is supposed to happen in the cotton mercerization process to improve its textile properties. It is given a thermal treatment with 15 to 25% concentrated sodium hydroxide solution, with the absence or presence of longitudinal force (Ottesen et al. 2019). Under these conditions, hydroxyl groups change the hydrogen bonding network, the cellulose swells, apparently producing a less crystalline structure with claims for a much greater free hydroxyl group content (Nomura et al. 2020). This form has better mechanical properties and improved dye adsorption of highly charged pigments. Mercerized cotton is often referred to as having the predominant presence of allomorph II (Mittal et al. 2011).

The classic view of allomorph II of having an antiparallel arrangement of simple glucan "ribbons" or single fibrils is based on X-ray diffraction techniques and neutron diffraction pattern interpretation. The antiparallel arrangement is a thermodynamic favored organization, easily rationalized in the regeneration process by a solubilization/precipitation process, as it is tough to occur in the designated wet-wet process. However, claims exist that this inversion does not need cellulose solubilization. This apparent lack of consistency may indicate the occurrence of another process. In some cases, mercerization does not involve the wet-wet process, but rather wet-dry process, however in the latter it is more difficult to visualize the parallel to antiparallel rearrangement. Other properties of mercerized cotton, such as its higher amorphous degree, the presence of free hydroxyl groups, combined with new studies in computational models, indicate that antiparallel chains do not easily explain the increase in volume, the 25 to 30% strength in properties mechanics and the presence of increased –OH groups. These results confer some uncertainty about exactly what the process and changes are involved in the mercerization.

In allomorph II, the classical view considers that some of the hydrogen bonds occur between the glucose units in different layers, which indicates that its organization cannot be described simply as a stacking of sheets as described for type I. The stacking is altered, and void formation is suggested

Table 2. Comparison of hydrogen bonding bonds and properties between accepted cellulose allomorphs.

Cellulose type	Intra single fibril bonds[1]	Inter single fibrils	Inter sheets	Cristal details	Elastic Modulus (GPa)	Example biomass	Notes
Allomorph Iα	2-6 and 3-5	2-6'	van der Walls	triclinic	138	Algae	
Allomorph Iβ	2-6 and 3-5	2-6'	van der Walls	monoclinic		Higher plants, cotton	
Allomorph II	3-5 and some 2-6		Van der Walls and hydrogen bonds	monoclinic	88		Anti-parallel single fibril
Allomorph III			3-5"	monoclinic	58–87		

1. Glucose carbon position notation: 2-6 hydrogen bond link refers to glucose belonging to a single fibril unit, 2'-6 or 2-6' between parallel single fibril, 2"-6 to a glucose hydroxyl in "different sheet" fibril.

by computational models, making Allomorph II, especially in hydrated form, the best reactive allomorph for enzymatic hydrolysis (Wada et al. 2010).

Allomorph III is obtained from I or II by treatment with amines or ammonia. This form shows a monoclinic crystal and is considered the most reactive (non-enzymatic) of all crystalline forms. The formation of 3-5" hydrogen bonds appear to be increased relatively to those in allomorph II.

Recent studies of Allomorph IV seem to consider it a variant of Alloform I (Kontturi 2015). Table 2 compiles the main characteristics of each of the cellulose allomorphs clearly identified to date.

2.1.2 Amorphous cellulose

Few studies have been carried out that attempt to determine the exact structure of amorphous cellulose (Bregado et al. 2019, Koizumi et al 2008). This reflects the difficulty of the problem, mainly due to the embedded nature of this phase between the crystalline zones, making their spectral information overlap with the latter. It is common sense that amorphous cellulose is less organized, filled with voids, with easier access to hydrolysis by enzymes or mineral acids. Amorphous cellulose data is easier to obtain from cellulose obtained by the regeneration/precipitation process using solvents. However, the structure determined in these samples may differ from the native form. Some authors using X-ray spectroscopy and neutron diffraction techniques suggest that mixed allomorphs may be present, with evidence of curved and twisted halves (Hans Finki et al. 1987). This means that different networks of hydrogen bonds coexist in the structure in adjacent positions, resulting in a shape with twists and folds leading to the formation of voids. Information from the mercerization process in which the effects of de-crystallization/crystallization are observed (Miura and Nakano 2015) indicate a profound change in the hydrogen bonding network of known allomorphs, with the net result of a smaller number of intramolecular hydrogen bonds (Begrado et al. 1997).

Computational models made of cellulose microfibrils with inter- and intramolecular hydrogen bonds characteristic of cellulose I were subject to change in their nature to others, characteristic of different allomorphs, and give rise to structures with curved and interlaced domains, similar to those interpreted by X-ray analysis. A model to exemplify a possible structure with this principle was built. In the model, segments of two ribbon cellulose fibrils with their ends exhibiting the characteristic 2-6'and 3-5' hydrogen bonds, typical of cellulose allomorph I, were constructed. In the central zone of the set of microfibrils, different hydrogen bonds were built, with the hydrogen bond 3-5' changed to atoms of different "layers" of cellulose, namely the bond type 3-5" (see Table 2 note). The structure was subsequently subjected to energy minimization. The result obtained is shown in Figure 4, in the form

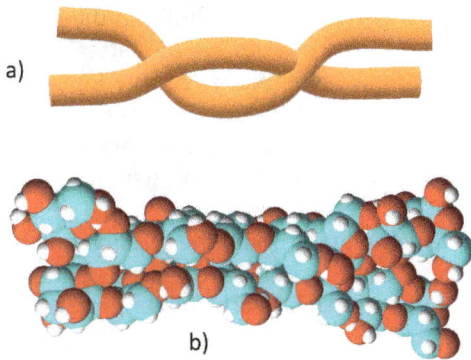

Figure 4. Multiple cellulose fibril stack with both ends with allomorph I characteristics (2-6'and 3-5' hydrogen bonds) and in the middle with a different allomorph type, only the 3-5" hydrogen bond.

of a chemical connection skeleton and in the three-dimensional version. In this model, the microfibril chain with central 3-5" bonds undergoes pronounced distortions from the normal fibrils positioned side by side at the ends to a twisted position about 90 degrees in the center, leaving some hydroxyl groups in positions where the formation of hydrogen bonds is unlikely and with the presence of twists. A sequence of multiple structures with this organization of hydrogen bonds will lead to the formation of voids and a less organized structure.

2.1.3 Gas diffusion and crystallinity studies

A complementary way to understand the dynamics of cellulose is through gas diffusion measurement modeling. Crystalline cellulose shows a compact spatial arrangement of glucose units that left none or very few voids in the structure. The study of the diffusivity of CO_2 in crystalline and amorphous cellulose (Mohammad et al. 2018) allows for a better understanding of the presence of voids and the compaction of polymeric chains together with the mobility of glucose atoms within the polymeric chain. The diffusion rate constant of carbon dioxide through crystalline cellulose was estimated between 3.33×10^{-9} and 3.20×10^{-6} cm^2.s^{-1}, and between 2.33×10^{-8} and $9,44 \times 10^{-6}$ cm^2.s^{-1} for amorphous cellulose (Mohammad et al. 2018). In the words of this author, "crystalline cellulose only undergoes reorganization (movement) of its hydroxyl and surface hydroxymethyl groups, while cellulose monomers (glucopyranose rings) linked to β(1,4) glycosidic bonds are fixed". The restricted movement of the cellulose monomers and the lack of space in the crystalline cellulose give rise to the restricted movement of the diffusing CO2 molecules.

In the amorphous regions, the glucose monomers appear to have a spatial position allowing diffusion rates around 1000 x higher, which means a less compact structure with less intensity in the hydrogen bonding network and with glucose oxygen's having a freer movement. These amorphous zones seem to be found along the length of the fibril both on the micro- and macrofibril scale, as seen in Figure 2 (see above). In the latter case, the possibility of intertwining with the other two biomass polymers—hemicelluloses and lignin—is suggested (Zhang et al. 2015b). Recent observations by the author favour the assumption. This association can serve as a form of structural reinforcement (Harris and DeBolt 2008) as well as making a way for the transport of water and nutrients.

Not all biomass has the same crystalline cellulose content. The crystalline predominance appears in the biomass of plants potentially submitted to strong shear forces such as cereals and other grasses. Several techniques are used to measure crystallinity, the determination of the X-ray diffraction pattern or FTIR spectrum absorption bands standout among the most used techniques. Its relative proportion is generally expressed in terms of the relative crystallinity index (RCI), other authors (Nelson et al. 1964) refer to the total crystallinity index (ICT). Table 3 presents some crystallinity data for different types of biomass (Harris and DeBolt 2008).

Table 3. Relative crystallinity value for several biomass examples, expressed as relative crystallinity index (RCI) and total crystallinity index (TCI).

Biomass	Botanical name	Tissue	RCI (average)	TCI	Notes/ref
Avicel			65.8		
			80.6		
Buddhist pine	*Podocarpus macrophyllus*	Leaf	33.9		Harris and DeBolt 2008
Arabidopsis	*Arabidopsis*	Whole plant	48.9		
Switch grass		Leaf	57.9		
Sweet Sorghum	*Sorghum bicolor Cretaceous*	Leaf	55.5		
Giant Miscanthus	*Miscanthus giganteus Cretaceous*	Leaf	57.9		
Sugar cane biomass	*Saccharum* spp.		50–8		Caliari et al. 2016
Wheat straw				1.13	Nelson et al. 1964
Sugar cane bagasse				1.18	
Rice straw				1.15	
Triticale straw				1.15	
Hydrolyzed wheat straw				1.05	
Cotton stalk			80.3		Li et al. 2019
Bamboo			78.6		
Algae cellulose			> 80.0		Konturi et al. 2015
Cotton linters			56.0–65.0		

The separation between crystalline and amorphous cellulose is also far from simple in both cellulose fibrils, but especially in macrofibrils. The most mentioned model suggests contiguous sequences of crystalline regions followed by amorphous regions; however, studies of SEM partially hydrolyzed samples of wheat straw cellulose suggest a different degree of twinning (in the simplest case) or even a mixture of intertwined amorphous and crystalline regions.

2.2 *Structure of hemicellulose*

Hemicellulose is a heterogeneous polymer as opposed to cellulose, making up about 25–45% of the gross weight of lignocellulosic biomass (see Table 1). As cellulose is the most abundant biopolymer on the planet, hemicellulose is, consequently, the second and first most abundant heterogeneous polymer, with different sugars entering its composition. Hemicellulose polymers have an average molecular weight less than about 30,000 Da, with a degree of polymerization (DP) of around 50–200 (Kontturi 2015), a value much lower than that achieved by cellulose (Anwarab et al. 2014). Other important aspects of the structure and composition of hemicellulose are the absence of crystalline structure and the presence of acetyl groups linked to the main polymer chain.

The location of hemicellulose is partially known and can be found mainly within the space of the cellulose macrofibrils, participating in the cell wall of the plant. Some authors suggest its existence also on the external side of macrofibril arrays (Rose 2003). Hemicellulosic sugars close to cellulose seem to suffer some effect induced from interference by the cellulose structure and the hydrogen bonding network, adopting a twist every 180° in a similar way to cellulose. At a greater distance, this interference does not seem to be effective, showing monomeric sugars in these zones a triple angle value with greater mobility and easier access to enzymes, as suggested by solid NMR studies (Simmons et al. 2016).

The most abundant sugar in hemicellulose are the five-carbon pentoses (C5), such as xylose and arabinose, connected through the β(1,4) glycoside linkage. The anomeric sugar pyranose C5 has the hydroxyl group in an equatorial position. For this reason, its backbone is linear, like that of cellulose. However, unlike the latter, hemicelluloses have a considerable degree of branching, with various types of sugars in the branches. They are there to avoid the formation of large crystalline regions as in the case of cellulose, thus making it amorphous, allowing the formation of voids that are organized in conductive vessels for movements of molecules such as water, simple sugars and nutrients (Adebayo and Martinez-Carrera 2014). The exact positioning of hemicelluloses, although not very well documented, seems to be important also for the flexibility of the cell wall. The connection between hemicellulose and cellulose appears to be stronger in the amorphous parts, where lignin may also be connected (see below). There is evidence that the binding of the hemicellulose to cellulose in biomass occurs predominantly through hydrogen bonding (Peng et al. 2011), unlike lignin, in which covalent bonds are often attributed to their binding to cellulose (Sung et al. 2011). This may explain the relatively easy extraction of hemicellulose from biomass with simple hot water and facilitates its access to chemical attack with suitable reagents.

The lack of the sixth carbon in the C5 monosaccharide rings allows its polymers to slide sideway more easily, in the occurrence of multiple chains. The same effect occurs in the case of the pair of polymers of polyethylene versus polypropylene, where in the last polymer the methyl groups obstruct the movement between the chains, giving a greater rigidity to the polypropylene. In the case of biomass, hemicellulose gives greater flexibility to the global arrangement of composite nature. The presence of some degree of acetylation in several hydroxyl groups is also observed; it can be a means of preventing the establishment of a high number of hydrogen bonds. In some types of biomass, binding with organic acids, such as phenolic acids (example cinnamic acids), has been reported (Arai et al. 2019).

The branching of the polysaccharide skeleton can block the entry of larger molecules, for example enzymes, which eventually could cleave the polymer, making hemicellulose highly complex to be enzymatically degraded (Puls et al. 1997). The presence of hexoses in hemicellulose can be vestigial or predominant depending on the subtype of hemicellulose. Hexoses such as galactose, glucose and mannose can be found from small amounts to predominant values.

The presence of glucuronic acid and galacturonic acids have been reported also as hemicellulose components (Arai et al. 2019). Other sugars such as rhamnose and fucose may also be present in small amounts. In some lignocellulosic biomass, hemicellulose fraction is made mainly by hexose C6 sugars–the galactomannan in softwoods–and show a considerable amount of mannose with minor contents of glucose and galactose (Baath et al. 2019).

This means that depending on the biomass source, different contents and types of hemicellulose are observed. Some types share C5 sugars, with β(1,4) glycosidic bonds, differing only in the composition of the side groups. Other types involve a completely different monosaccharide monomer backbone. These different types of skeleton lead to the division of hemicellulose into subgroups, depending on the sugars and the arrangement of the known building blocks of the primary structure. Considering the primary structure of the sugar polymer, hemicellulose can be subdivided into five categories: xylan, glucuronoxylan, arabinoxylan, glucomannan and xyloglucan (Xinxin et al. 2019). Other authors simplify the classification by considering the five preceding categories represented mainly by three divisions. These are three main subgroups of hemicelluloses that have a β(1,4)-D-pyranose residue structure, namely xylan, xyloglucans and mannans, see Figure 5 (Naidu et al. 2018).

The composition of hemicellulose differs between plant species and therefore in biomass. For example, straw and grasses are mainly composed

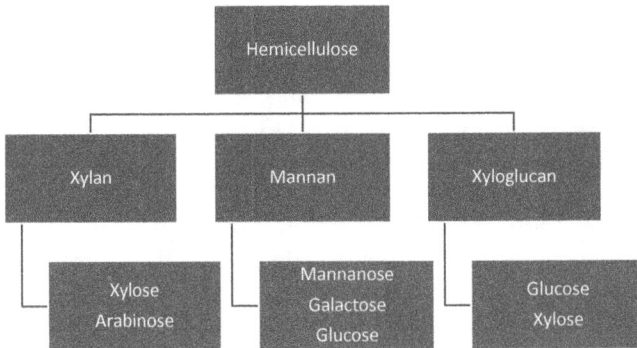

Figure 5. Biomass hemicellulose levels of organization with the main monosaccharides found in its composition.

of xylan, which is easily extracted with aqueous acid or alkaline solutions, while softwoods contain mainly glucomannan which, unlike xylan, requires a strong alkaline solution for its extraction. Xylan is the most abundant type of hemicellulose in the plant cell wall, with xylose as the main monomer (Linder et al. 2003).

2.2.1 Hemicellulose biosynthesis

The problem of xylan organization in the secondary plant wall remains unclear, suggestions have been made that the Golgi apparatus in the cell is responsible for its synthesis (Meents et al. 2019). In a subsequent step, it is added to the outer layers of cellulose microfibrils formed in the CESA rosettes (Linder et al. 2003). Under this view, lignin will be the only main biomass component to be polymerized in the cell wall by diffusion of monomer *in situ*, that is, in the final location. Lignin biosynthesis is believed to be done by the enzymes lignin peroxidase, manganese peroxidase and laccase. The problems with this model are the microfibrils are organized in arrangements forming macrofibrils that do not appear to have any hemicellulose incorporated in its inside. A possible rearrangement process can been visioned, with the expulsion of the low-DP hemicellulose fibrils to the outside, by the junction of multiple simple cellulose fibers to form the crystalline cellulose, however, doubts arise when taking into account the great degree of branching of hemicellulose and its effects in obstructing this process.

A mechanism involving both the biosynthesis of lignin and hemicellulose from their monomers, could explain the interplay of oligoxylanoside chains with lignin. The major problem with this model is that it does not find evidence of the existence of xylan synthesis enzymes. This part remains relatively obscure in the synthesis of lignocellulosic biomass. However, in the outer layer of cellulose macrofibrils, there is undoubtedly a type of saccharide with reactivity similar to hemicellulose.

2.2.2 *Xylan*

Xylan is the main subtype of hemicellulose found in hardwood and herbaceous plants. Xylan consists of the β(1,4)-D-xylanpyranose structure containing different monosaccharides as side substitutes, such as glucuronic acids, often showing some degree of acetylation in its 2 or 3 position, or both. Other side chains observed are α(1,2)-D-glucuronic acid, sometimes with 4-O-methylation and β(1,3)-L-arabinofuranosyl groups (see Figure 6). Branching prevents crystallization and facilitates the degree of hydration (Arai et al. 2019). Glucuronic acids are known to allow the formation of cross-links in polymers with the coordination of divalent ions such as Ca2+ or Mg2+. A characteristic consequence of having different side chains is to transform attacks of difficult microbial enzymes in both the sugars of the structure and the side substituents.

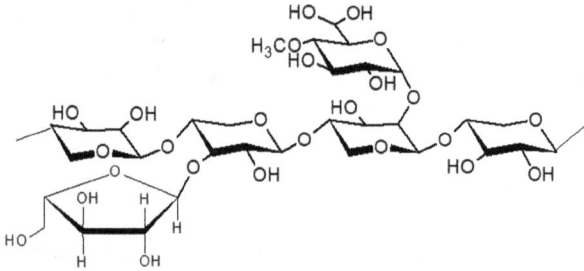

Figure 6. Typical xylan hemicellulose structure, showing the skeleton of the C5 structure replaced with glucuronic acid (at the top) and an arabinose residue (at the bottom).

2.2.3 *Xyloglucan*

Xyloglucan shares the same C6 skeleton with cellulose, but has considerably more complex side chains, from simple C5 sugar to different length C5 sugars linked to hexose oligomers of different size, predominantly linked in 2 or 6 positions of the C6 sugar skeleton. An example of the repetition structure of xyloglucan is shown in Figure 7.

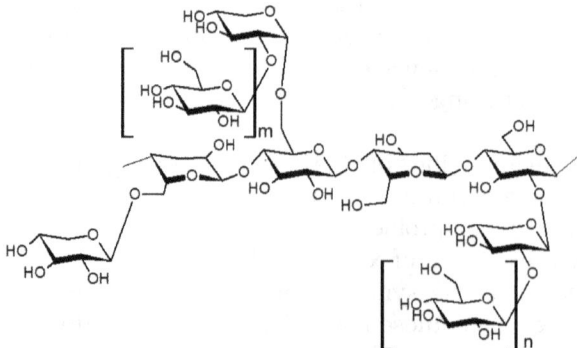

Figure 7. Accepted structure for glucosylxylan hemicellulose. The side chains may in some cases have the xylose-galactose-fucose trisaccharide.

2.2.4 *Mannan*

The mannan hemicellulose type consists in a polymeric backbone where mannose is the predominant sugar (about 65%). The main chain shows branching points, occurring primarily by a single galactose sugar residue connected through β(1,6) linkage to mannose, in lesser quantity side chain sugars, such as arabinose or different types of disaccharides can be present. Acetate groups are present in some backbone mannose sugars, mainly in the sixth mannose carbon. Guargum has a structure related to mannan (Furquim Da Cruz 2013), the same structure also occurs in coffee beans (Bekedam 2008), see Figure 8 for the general structure. Several mannan variants are found in biomass with different structure modifications in the galactose/glucose structure copolymer (Konturi et al. 2015).

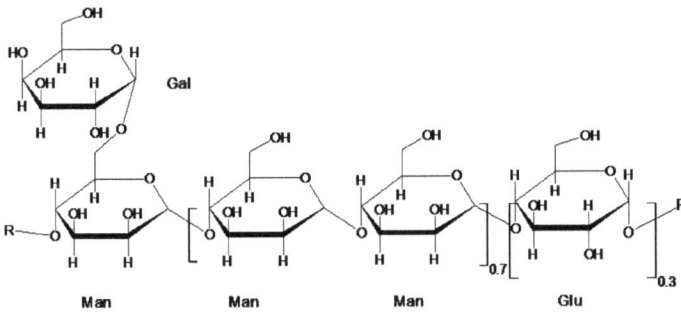

Figure 8. General mannan structure, with β(1,4)-D-mannose copolymers with partial content of D-glucose portions, in a ratio of approximately 7/3 and with galactose ramifications.

2.3 *Lignin*

In biomass, in addition to cellulose and hemicellulose, lignin is the third polymeric material found. It has an essentially hydrophobic nature and is difficult to extract from the biomass. Lignin can be found in proportions that reach values in the range of 10–30% (w/w) (see Table 1), that is, it constitutes the smallest fraction of the three main components of biomass. Due to its structure which is rich in aromatic rings related to phenols, it constitutes the most abundant polymer found in Nature with an essentially aromatic nature. The values in Table 1 show that herbaceous plants have the lowest lignin content, while softwoods have the highest content (Suhas et al. 2007).

Lignin is an amorphous substance encountered in the space between the phases of cellulose macrofibrils, and is found together with hemicellulose forming a porous and flexible domain responsible for the transport of water, nutrients and some enzymes. It also has connection points with cellulose and in this sense it can contribute to the mechanical integrity of the composite arrangement of biomass.

In the biosynthesis of lignin, vinyl-phenolic monomers are polymerized catalyzed by different enzymes, the most important of which are manganese peroxidase and lignin peroxidase (Varejão 2001). The precursor alcohols are p-coumaroil, coniferyl and synaphyl (see Figure 9). The possibility of formation of some cross-links in lignin provides mechanical resistance of the cell wall and protection of biomass against attack by microorganisms (Saini et al. 2015a).

Details of the location of lignin in the biomass suggest its proximity to hemicellulose with probable connections or entanglement with cellulose fibers, which will be more likely to happen in the amorphous regions; some authors refer to the presence of covalent bonds between lignin and cellulose (see above). The proximity of lignin to hemicellulose is supported by C13 NMR magnetic coupling studies (Talebi et al. 2019). The removal of lignin by selective methods is known to lead to the structural collapse of cellulose and hemicellulose (Ding et al. 2018).Therefore, lignin will be an essential component to guarantee the separation of sugar polymers that confer mechanical resistance and allow porosity in the hemicellulosic/lignin domain. Depending on the source, lignin has a slightly different structure. Lignin from biomass with long fibers is richer in coniferyl alcohol (approximately 90%), while grass lignin is richer in coumaric fractions, and hardwoods are abundant in sinapyl alcohols (Yong et al. 2017). The use of sub and supercritical fluid has been used to isolate lignin for extraction from cellulosic biomass and are used as a precursor for the production of carbon fiber adjuvant, in a content of about 15% maximum. Other interesting and poorly studied properties is the fact that it has a strong antioxidant activity, which can be expected considering the presence of a large number of phenolic groups (Mahmood et al. 2018). Like other phenolic substances, lignin has properties to capture and stabilize free radicals, thus providing resistance against oxidative stress to cell walls. The classic view of the structure of lignin describes it as a randomly polymerized phenolic "resin", with the presence of a large number of cross-links. The structure would be rich in several types of chemical bonds, as expected from the radical polymerization of three different types of vinyl monomers. The nature of cross-linking involves reactions between phenolic monomers, with formation of carbon-carbon, C-C and also ether bonds C-O-C (Boeriu et al. 2004). Ferulic acid may be present, linked covalently through the ester bond, eventually forming bonds with hemicellulose (Luo et al. 2019).

In recent years, efforts have been made to better understand the structure of lignin. The new data suggest a structure with the presence of a linear chain

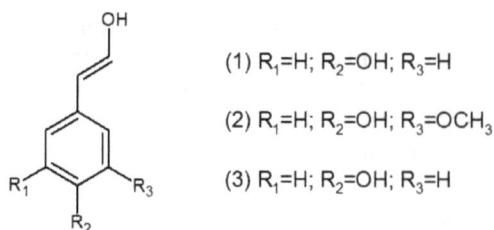

(1) R_1=H; R_2=OH; R_3=H

(2) R_1=H; R_2=OH; R_3=OCH$_3$

(3) R_1=H; R_2=OH; R_3=H

Figure 9. Lignin alcohol precursors: (1) p-coumaroyl, (2) coniferyl and (3) sinapyl (Suhas et al. 2007).

with twists, U-folds, in some places Y-derivations and a lesser presence of cross-links (Yao et al. 2017).

Table 4 shows some of the main real phenylpropane connections recently identified and verified (Ralph et al. 2019). In addition to the variety of bonds, two types of bonds can be emphasized—one involving a skeleton of alkyl chains connecting the phenyl rings, the other, alkyl-ether bonds connecting the aromatic rings. In all cases, these different types of structures always contain benzilic carbon, which, as will be seen, defines the characteristic reactivity of lignin.

Table 4. Lignin phenyl propanoid linkage structure example, the number of connections in the lignin backbone, and the corresponding number of benzylic positions.

Linkage designation	Type	Number of connections in polymer chain	Structure	Number of benzylic positions
β-O4-β ether	Alkyl ether (a type, in structure, in bold)	2		1
β-5-phenyl-coumaran	Both cyclic alkyl ether (a) and cyclic alkyl carbon chain (b, in bold)	2		"2"
5-5/4-O-β dibenzo-dioxocin (DBDOX)	Cyclic alkyl ether (a type, in structure, in bold); biphenyl-like C-C bond (b, in bold)	2/3		1

2.4 Minority compounds

As mentioned in the section on lignocellulosic biomass, plant cell walls are composed of three main compounds: cellulose, hemicellulose and lignin. However, other components can be found in small quantities (White et al. 2020, Lema-Ruminska et al. 2019). Pectin is most often present in biomass and consists of a polymer chain of α(1,4) galacturonic acid with frequent esterification of carboxylic acid by methyl groups. All biomass, as expected, contains other components. Protein is found in all types of biomass, as it is the raw material for enzymatic machinery capable of synthesizing this organized life form. Water and mineral salts are always present at different levels depending on the type of biomass. Other compounds, such as secondary metabolites, may be present, but they do not participate significantly in the formation of the plant's cell wall structure. Biomass can contain many other substances, one example is wax, most biomass is coated with these very hydrophobic materials that are involved in inhibiting dehydration, in some cases in amounts of 2–3% (w/w). Pine wood, for example, is rich in terpenes, mainly monoterpenes α and β-pinene together with diterpenes such as abietic acid as the main. The seeds generally have a high content of lipids, in addition to a reserve of starch.

3. Proposed structure for biomass

Taking into account the most recent observations made on different types of biomass with updated results published in the literature described in the previous sections, an updated proposal model for the organization of the three main lignocellulosic components is shown in Figure 10.

Lignocellulosic composite materials appear to be formed essentially from two different domains, see Figure 10. The first domain is cellulose fiber. It arranges single cellulose microfibril units (primary cellulose) into multiple stacks, the microfibrils, which may contain numbers from hundreds to thousands of individual cellulose single fibers. In these it can be found two distinct regions, one with a high degree of crystallinity and constant allomorph type and others with a greater ease of acid hydrolysis, of a more amorphous nature and may be either a different allomorphic type or mixtures of different allomorphs along the chain. There was never any evidence that the sugar composition of the crystalline and amorphous regions was different. These microfibrils are hierarchically organized into superstructures giving rise to secondary macrofibrils, in the case of wheat straw, arranged in super-arrays resulting in micrometric size already visible under the microscope, ranging from 200 to 600 μm. The macrofibrils appear to have a coating made of a material other than cellulose, probably a layer of pentose oligomers similar in composition to the surrounding hemicellulose, but which are apparently affected by the effects of the cellulose hydrogen bonded network, thus changing its organization. This interference will arise from the compatibility of hydrogen bonding between the two polymers

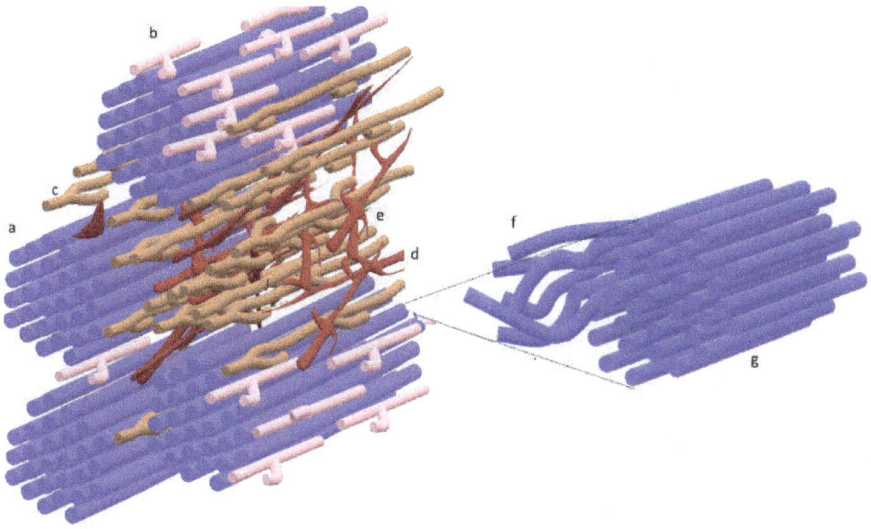

(a)

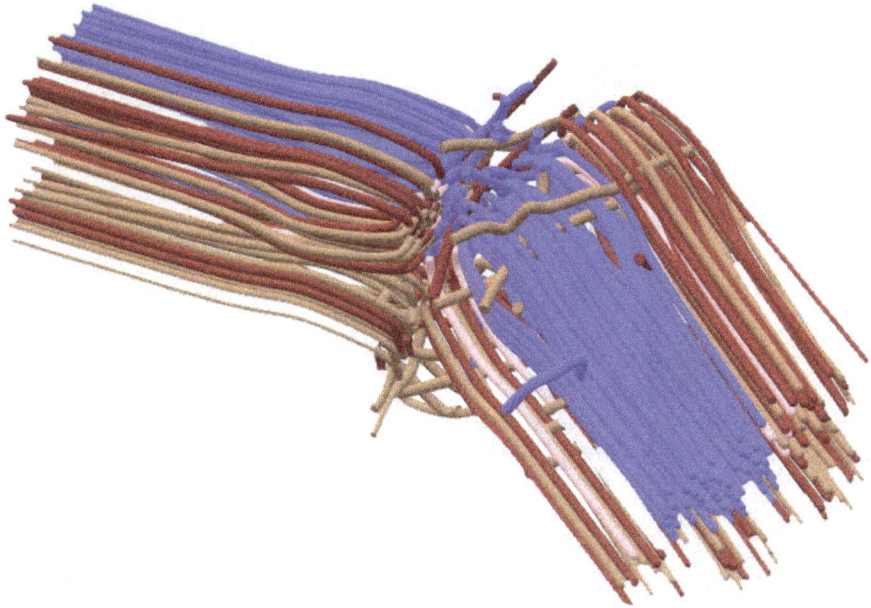

(b)

Figure 10 contd. ...

... Figure 10 contd.

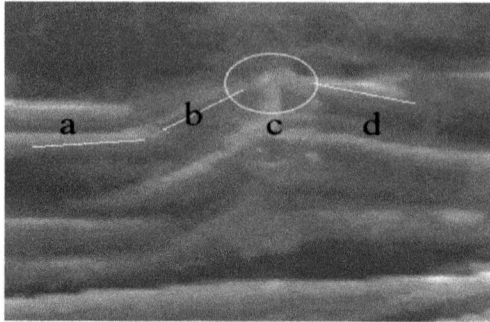

(c)

Figure 10. (a) Simplified model of a biomass structure segment based on recent observations. (a) Cellulose macrofibrils (a in blue) are organized in super arrays containing tens to hundreds. Both super-arrays or macrofibrils appear to be coated with a different polymer, probably of the hemicellulose type (b in pink). Each macrofibril is made by joining a large number of microfibrils (g in blue), themselves formed by assembling a large number of individual cellulose fibrils. Microfibrils contain crystalline (g) and amorphous (f in blue) regions. This set constitutes the cellulose domain. The cellulose domain is immersed in a hemicellulose/lignin domain (respectively, c in light brown and d, in dark brown) forming a porous, hydrophilic material, where the voids allow the movement of liquid solutions. (b) A simplified model of the partial cut of the amorphous macrofibril zone. The chains of crystalline cellulose alter its allomorphs and change direction, creating voids where connections with hemicellulose and lignin occur (the central region). A SEM image of such a zone is shown, from a sample of wheat straw, in (c), with removed hemicellulose and partial lignin fractions. A macrofibril following the direction from a to b, appears as an unorganized (circled) amorphous zone. The macrofibril continues in a different direction in d. The same effect is repeated twice in the fiber positioned below the assigned one.

of different nature, resulting in an inter-phase that is very reactive to the presence of acids (assigned in rose in Figure 10a).

The "amorphous" regions of cellulose are found to occur both on the scale of microfibrils and macrofibrils (see Figure 10).

The second domain has a nature composed of hemicellulose polymers (the light brown color in Figure 10) intertwined with polymeric lignin chains (the dark brown color in Figure 10). This structure is rich in empty spaces resulting in a structure of porous vessels with hydrophilic characteristics. Supporting this model are recent studies of coupling solid state NMR C13, suggesting that the contact between hemicellulose and cellulose is not intense (Wang and Hong 2016). This second domain is built in such a way that it allows the contact of the cellulose fiber arrangements with it at some points, in order to allow the transport of water, building blocks and nutrients to the two domains. At this points, the two domains seem to merge, being locals where the closest proximity of the three main biomass polymers of the cell wall occurs. A structure resembling a twisted and distorted triple helix is formed in these regions, which gives it an essentially amorphous nature (Simmons et al. 2016). These regions have more intense chemical bonds with hemicellulose and in a stronger way lignin which can contribute to the mechanical properties of the material, while allowing the movement of water and nutrients. A model of such a region is shown in Figure 10b. The evidence for such a region comes from different SEM image observations of biomass and also derives from the closed form of the cells, together with the need to carry out the cellulose fibril elongation process, usually in a specific direction, by the joint action of a large number of cells. If we consider the cell's position in relation to space, its positions should deviate slightly from its neighbors and then from linearity, making the growing polymer only partially successful in continuing the existing growing sugar chains. This led to (i) the change in the direction of the crystalline primary cellulose chains; (ii) the impossibility of giving continuity to some of the polymers already formed, explaining the fringed cellulose structures as frequently observed in SEM images, resulting in "loose" ends of the polymer matrix.

The exact arrangement of the two domains depends on the type of plant and the local tissue. Apical development occurs in grasses, which have an almost perfect straight elongation. Other types of biomass show a high cross pattern of cellulose macrofibrils with the surrounding hemicellulose/lignin domain, see Figure 11.

It can be inferred that the composite nature of terrestrial biomass, joining crystalline fibrous parts with an adjacent domain with polymers with some degree of cross-linking, obstructs the passage of enzymes and confers recalcitrant properties to biomass in relation to physical, chemical or enzymatic attack, making it difficult to deconstruct lignocellulosic materials. It constitutes a polysaccharide organization opposite to the polymers of macroalgae, where its structure was assembled by Nature in such a way that other factors, such as water retention is prevalent and prevents dehydration, resulting in global arrangements that are easier to "dismantle" and in which

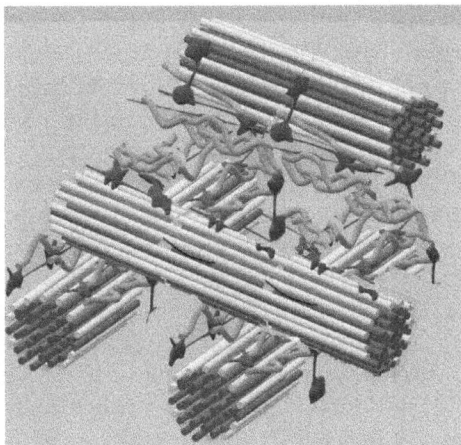

Figure 11. Transversal arrangement of macrofibrils found in types of biomass, such as different types of wood.

conversion of polymers to sugars can be achieved with great ease (Varejão and Nazaré 2017). uThe strong structure of terrestrial biomass has been the main disadvantage of its use for the production of cellulosic ethanol (Saini et al. 2015b) however these properties can be used to obtain other biomaterials with very strong mechanical properties, which can be useful in various fields and products for Humanity.

4. Disassembly of lignocellulosic materials

4.1 Physical treatments

As with any other analytical processing method the access of reagents to the analytes to be chemically modified is vital. Regarding biomass, it appears in different shapes and sizes, from small dimensions in one axe as in cereal straws, to a much larger dimensions as in tree trunks. The conversion of biomass into materials with small particle size or partial defibrillation facilitates any process designed to recover the constituent components. In this sense, pre-treatments such as crushing, grinding, chipping, and steam explosion help to facilitate physical access to reagents. The grinding is a process with high energy consumption and expensive, which makes it undesirable, however, the steam explosion method can be very effective while consuming little energy. In this process, ammonia or water vapor is pressurized to diffuse through the biomass pores, after equilibrium is reached, at pressures of around 50–60 atm. A sudden release of pressure causes the compressed gases to a rapid escape, creating an "explosion of biomass" from inside, which largely defibrillates it resulting in a material with characteristics similar to "cotton". However, as discussed earlier with the diffusion of carbon dioxide in cellulose, crystalline cellulose is not expected to be much affected.

4.2 Chemical treatments

The complex structural characteristics of biomass make the extraction of each of its components particularly difficult. However, their recovery is a very desirable objective, since from them a wide range of products with market demand could be prepared. Unfortunately, the composite nature of the material makes this objective difficult, and require to evolve techniques that are contrary to structural recalcitrance.

A disassembly strategy that has been particularly studied is the attempt to saccharify the biomass to obtain as much sugars as possible, leaving lignin as a residue. Polysaccharides represent 70 to 80% of the biomass composition in terms of dry matter, all of which are potentially fermentable. The production of ethanol from biomass residues is a very desirable goal. However, hemicellulose-derived sugars are mainly C5-based sugars, such as xylose and arabinose, whereas cellulose decomposes to glucose. In terms of fermentation capacity, glucose is easy to ferment, unlike sugars in C5 (Ankit et al. 2012, Singla et al. 2012).

The van Soest method (van Soest 1998), which allows the determination of the nutritional value of plant foods for ruminants, is one of the oldest processes for fractioning biomass into its main constituents. The method starts by isolating two fractions, from dry biomass; the fraction of neutral detergent (NDF) and the fraction of acid detergent (ADF). The first essentially removes lipids, soluble resins, waxes, free sugars and some minerals, corresponding approximately to the total content in the biomass of the sum of hemicellulose, cellulose and lignin. Treatment with acidic detergent mainly removes hemicellulose, which is relatively easy to hydrolyze with the use of diluted acids in the presence of detergents. From the ADF fraction (approximately cellulose plus lignin) treatment with hot permanganate, lignin is depolymerized and dissolve, leaving cellulose, the same fraction treated with sulfuric acid at room temperature and in a concentration of 72% (v/v), for 24 h, dissolve completely cellulose trough hydrolysis. This can be removed simply by washing with water. With this analytical scheme, the contents of the three main constituents of biomass can be determined, together with the mineral content. From these composition data, the ruminant nutritional value and digestibility can be derived.

During the last decades a great effort of investigation in processes that effectively carried out biomass disassembly has been studied, highlighting three main methods:

– the method of using diluted acid,
– the method of using concentrated acid,
– enzymatic methods.

The vast majority of research carried out was based on the production of biofuels, namely ethanol (cellulosic ethanol), which meant trying to maximize the yield of fermentable sugars from biomass.

4.2.1 The diluted acid method

One of the strategies studied over time to saccharify the biomass is the treatment with diluted acid, in concentrations below 5% and under relatively high temperatures, in the range of 80°C to about 240°C, in the last case using pressurized vessels. Under these conditions, hemicellulose is easily removed and quickly hydrolyzed to monomers, however cellulose is only partially converted to glucose, predictably originating from the amorphous part. The literature provides many works with these conditions (Zhao et al. 2015), sometimes with claims of obtaining a 90% yield in the theoretical conversion of biomass into sugars. However, some difficulty in depolymerizing crystalline cellulose is evident. The process never passed the laboratory or pilot phase to constitute a selection method for the production of sugars or ethanol from biomass.

The main reason for the low success of this technique lies in the fact that the saccharification of monosaccharides occurs sequentially—the first sugars to be obtained are those corresponding to the fraction of hemicellulose, subsequently the amorphous cellulose begins to be hydrolyzed, leaving a residue that slowly is attacked, the crystalline cellulose. The big problem is that sugars that have already been depolymerized, especially pentoses, cannot withstand the thermal conditions of the process. The last and part of the hexoses already formed are converted into furfuraldehyde (from C5 sugars) or 2-hydroxymethylfurfural (from C6 sugars). These substances are produced in levels that can reach values of about 3–4%, depending on the degree of severity of treatment, which depends on the time and temperature to which the biomass is subjected. Furfurals are known to be strong yeast fermentation inhibitors, making subsequent easy conversion to ethanol difficult. The microorganism most capable of fermenting sugars in ethanol is the yeast *Saccharomyces cerevisiae*, this and other yeasts are strongly inhibited with the presence of these substances. A workaround for this problem involves removing furfuraldehydes from the medium, with methods such as adding lime Ca(OH)2, activated carbon or using steam distillation purification. A two-stage method was developed, isolating C5 sugars in the first phase under milder hydrolysis conditions, and in a second phase trying to obtain glucose at higher temperatures.

In general terms, the standard result generally obtained in this process is either low saccharification yield with low furfural content or good saccharification yield with the presence of a high furfural content. Both results make the use of these conditions unfeasible if the ultimate goal is ethanol production by fermentation.

4.2.2 Concentrated acid method

The van Soest method indicates that cellulose can be dissolved in 72% sulfuric acid at room temperature. Any researcher who works with concentrated sulfuric acid in laboratory has a collection of clothing with various types of holes—sometimes some that appear only after the pieces of clothing have gone through multiple cycles in the washing machine. Glucose is known to be stable in a solution of concentrated acids at low temperatures. The same is not true for hemicellulose pentoses, which are very reactive with concentrated acids producing furfural and other subsequent condensation products, the latter responsible for the darkening of solutions. A great advantage of using concentrated acid is that they make it possible to dissolve cellulose crystals, to allow subsequent hydrolysis to glucose without appreciable decomposition at low temperatures, close to zero degrees Celsius. Both sulfuric and phosphoric acid are particularly effective. In contrast, the process requires prior removal of hemicelluloses—a problem shared with the method that uses enzyme treatment—see below. The presence or absence of lignin in the process of saccharifying the sugar polymers of biomass has been studied by several authors (Geng et al. 2019). Partial removal of lignin can be beneficial to saccharification, as seen in results of saccharification of biomass previously treated with brown rot fungus ligninolytic enzymes (Deswal et al. 2014).

However, its majority removal can cause the collapse of the hemicellulose/lignin domain, with the aggregation of cellulose macrofibrils, resulting in a dense material that prevents the access of acids, making the saccharification process more difficult.

4.2.3 Use of enzymes

That biomass can be dismantled quickly and effectively by enzyme cocktails has been known for some time from the digestive characteristics of termites (Kundu et al. 2019), that feed on some types of wood and excrete lignin almost in a pure state in a digestion period of about six hours.

The carbon cycle, in its final stages, involves the hydrolysis of hemicellulose and cellulose polymers through enzymatic depolymerization. For this objective to be successful, a multiple set of enzymes is required—delignification using enzymes such as peroxidases together with a complex mixture of specific C5 and C6 cellulases, for the hydrolysis of hemicelluloses and cellulose, respectively. Glycosidases are necessary for the final conversion of oligosaccharides to glucose. Many other types of life, microorganisms such as fungi and bacteria, can depolymerize the biomass to obtain food in this way in order to live. The metabolic pathways of microorganisms involve possibilities of metabolites exchange, in addition to sugars, to the host organism with mutual benefit. Several microorganisms present in the digestive system of termite flora seem to produce the complete set of enzymes necessary to consume all the C5 and C6 sugars present in the biomass. Enzymes are described as a mixture of cellulases, along with some cellobiolases. A parallel mixture of enzymes for saccharification of hemicellulose is also reported. The relative amount of each enzyme depends on the organism and the type of biomass to be processed.

In each of the two sets of enzymes, at least three groups of enzymes were identified, the endocellulases that are capable of superficially de-crystallizing cellulose, making cuts in the microfibrils that leave two loose ends; exocellulases that make random cuts in the loose fibers that have been released, giving rise to oligomers of different dimensions and glycosidases that finally convert the oligosaccharides into glucose. These enzymes are produced inside the cell of microorganisms and brought out as extracellular enzymes. Aquatic microorganisms use strategies in order to create conditions so that the enzymes or sugars already obtained do not diffuse in the medium, via the construction of carbohydrate gel capsules or closed cellular aggregates. The industrial production of cellulase enzyme mixtures with very high specific activities is currently available on the market and commercialized by different brands. The production of these enzymes should be considered a major technological achievement, requiring extensive knowledge and skills in different areas, which involve the selection of microorganism species, obtaining new variants produced by mutagenesis/selection sequences and more recently, by altering genome through genetic engineering techniques. Its availability allows useful end uses in industries such as textile treatment, household items such as textile softeners, etc. The attempt to use these enzymatic systems in the processing of biomass, aiming at ethanol production, has also been extensively studied and constitutes one of the main branches of research in the attempt to convert biomass into ethanol (Lynd et al. 2017). The successful use of enzymes can be seen in the industry in a multitude of usages. Enzymes, however, are expensive and difficult to obtain in significant quantities, most often at the expense of a lengthy and laborious multi-step process. Additionally, being biological entities, they lose activity in a relatively short time, with the use, degradation by oxidation, etc. Their use is very competitive, especially when they can be immobilized in a matrix that retains them, with activity intact and at the same time allows the substrate, the reaction product and the necessary cofactors permeate through the matrix. These manipulations allow the same enzyme to be reused a very high number of times (number of turnover).

When trying to dismantle biomass using enzymes, circumstances arise that are difficult to circumvent. On the one hand, enzyme immobilization is impossible to be applied in the case of biomass degradation, since the substrate in this case is a superstructure. This means that the enzymes have to move "individually" to the substrate, and their performance is difficult at different levels; hemicellulose and cellulose polymers form an intricate structure with complex details and precise shaped channels with a design that makes it intricate for enzymes to enter. Nature prevents its passage as a defense method, as seen in the high degree of branching of hemicellulose. At the same time, lignin is always present and has excellent adsorption properties, capable of retaining enzymes in particular. For example, lignin is known to be a strong cellulase adsorbent (Saini et al. 2015a). Physical treatments can work around these problems, but crystalline cellulose is not affected by these treatments.

The recovery of enzymes at the end of the catalytic process is also an issue with a low probability of success. It is not feasible to use expensive enzymes that are used only once to obtain products that are certainly worth less than the enzymes themselves. Research in recent decades has focused on this strategy, and since the enzymes are to be used only, the solution has been to increase its activity. Several modifications or different versions (see below), using enzymes millions of times more active than their original precursor, with the native inhibition mechanisms erased, have been the proposed solution.

A possible exception in the use of immobilized enzymes in the successful treatment of biomass may occur with glucosidases, which, being able to hydrolyze small oligosaccharides which are relatively small molecules that can diffuse from biomass and be processed superficially, approaching the appropriate characteristics for hydrolysis with immobilized enzymes.

All of these problems mean that the use of enzymes—at least until now— has led to the failure of the so desirable "cellulosic ethanol" (Lynd et al. 2017).

5. Biomass saccharification/fermentation process

Using the above methods for biomass saccharification and eventual subsequent fermentation, several techniques at the level of unit operations were attempted by the researchers. A description of the main variants is provided.

5.1 Simultaneous saccharification and fermentation (SSF)

Simultaneous saccharification and fermentation (SSF) refers to the processes in which the sugar polymer hydrolysis process is carried out in the same container in the presence of yeasts capable of converting them to ethanol. It is assumed that glycosidic bonds are broken by acidic or enzymatic activity, while the conversion of sugar into alcohol, in most cases, is done with the yeast *Saccharomyces cerevisiae*. One of the advantages of the medium is that the hydrolyzed monosaccharides can be quickly directed to fermentation, without being subject to degradation. The production of ethanol from starch materials uses this process with relative ease and efficiency, constituting a well-stabilized technological process (Bottger and Sudekum 2018). However, in this case, it is easier to break glycosidic bonds and the sugar obtained is fermentable. Many tests have been done to apply this strategy in the case of fermentation of lignocellulosic materials. SSF has a number of advantages and problems—it is a simpler process that requires less hardware, since it can be done in a single reactor, but it turns out that the simultaneous presence of microorganisms capable of producing extracellular enzymes capable of breaking the bond glycoside with others that metabolize the same substrate create mutual problems that lead to inefficiencies in the process. Enzymatic inhibition is one, generally nature makes enzymes effective only in a narrow range of glucose concentration and is strongly inhibited at higher concentrations, in this case both glucose and ethanol act as the inhibitory species. Genetic engineering has tried to solve the problem by introducing or removing genes that add or remove, as example enzyme inhibition, and lead to the modification of metabolic pathways, sometimes with unexpected results in a metabolically unrelated process, but necessary for good general activity. Ethanol concentration is another issue not resolved—as the ethanol concentration increases, organisms that produce cellulolytic enzymes cannot withstand and start to fail, generally at ethanol levels above 5–6%; therefore, in more advanced stages of the process, the lack of sugars interrupts the process. In this process, the fate of C5 sugars is always uncertain. It is not a good strategy to forget the general structure and composition of the biomass, and about 30% of the biomass weight are pentoses, if they are saccharified a microorganism must be able to convert them into ethanol. The yeast *Saccharomyces cerevisiae* is not able to do this. The use of different yeasts is not yet a possibility, since no other has been found capable of withstanding concentrations of ethanol above 5–6% (v/v).

5.2 *Separate hydrolysis and fermentation (SHF)*

The SHF process is one that tries to overcome the disadvantages of the SSF discussed above. It corresponds to the first stage of saccharification of biomass in a separate vessel, and the subsequent fermentation of the sugars obtained in a second vessel. This process can solve part of the problems associated with the SSF process. However, it is more demanding in hardware complexity, requiring the presence of at least two to three separate reactors. Saccharification produces a solution containing C5 and C6 sugars, so fermentation poses a problem, as yeast *Saccharomyces cerevisiae* is not able to convert C5 sugars. The enzymes that are lacking in its genome for the possible metabolic pathway are the xylulose reductase and xylulose isomerase, genes that were subjected to the insertion into yeast plasmids in genetically modified variants. But even in these modified microorganisms, the use of glucose is preferred over xylose (Vilela et al. 2015). Unfortunately, so far, the interference of cross-glucose/xylose inhibition in metabolic enzyme pathways has not produced a yeast capable of efficiently fermenting both sugars in C5 and C6 (Inokuma et al. 2017). This topic is an active and intense focus of research that may in the future provide better solutions, apparently their immobilization can benefit their activity (Antunes et al. 2019). Other yeasts with xylose fermentation capacity generally have low tolerance to ethanol (maximum concentration < 6–7% v/v) making its use unviable if ethanol is the intended product. Although microorganisms capable of converting both C5 and C6 sugars do not exist today, the best strategy will be to first remove C5 under labile conditions and convert it into other useful bioproducts. The remaining biomass must undergo a process of de-crystallization followed by saccharification techniques, by acid, enzyme, or a mixture of both (Guo et al. 2020).

Different variants of this methodology were tested, such as simultaneous saccharification and co-fermentation (SSCF), with claims for ethanol yields greater than 80% of the theoretical calculation, for example, of wheat straw residues (Erdei et al. 2012).

5.3 *Consolidated Bioprocessing System (CBP)*

Recently, the focus on research in the area of obtaining ethanol from biomass has frequently focused on the so-called Consolidated Bioprocessing System (CBP) (Jouzani et al. 2015). The method follows a logical path–if biomass is a diversified and complex natural material that requires a large number of enzymes for its complete disassembly, the idea is to mix all the enzymes and/or the organisms capable of producing the substances necessary for its depolymerization.

The joint fermentation of pentoses together with hexoses is necessary in an industrial perspective to make the price of ethanol competitive. If possible, this C5 and C6 sugar fermentation of sugars should be done by the SSF process, to simplify and improve the ethanol yield. Much research has been done on this strategy, but so far with poor results. An already complex problem seems to become much more complex. This process involves the addition of several types of enzymes/microorganisms capable of carrying out the necessary multiple steps, but problems of cross enzyme inhibition, incompatibility of microorganisms, arise. Genetic engineering strategies to design more suitable microorganisms were used. Zhang et al. (2015) showed that a mixture of C5/C6 co-fermentation by means of genetic engineering microorganisms, *S. cerevisiae* and *E. coli* were able to convert biomass into ethanol in low yields. Recent results on the immobilization of microorganisms and genetically modified enzymes seem to pave the way to reduce its complicated set of dependencies and, thus, optimize its capacity (Antunes et al. 2019). The discovery and manipulation of cell transporters has enabled these new results that look promising. The manipulation of the type of transporter can also increase the capabilities of *S. Cerevisiae* to improve the direct use of oligosaccharides (Alves et al. 2008). This can in the future make the PBC process possible. A recent comparison of the state of the art of methodologies for ethanol derived from biomass referred to the delay in the search for solutions (Rastogi et al. 2018) keeping the processes very low yields and making the industrial process unfeasible. At some point, this solution can give results that are not yet defined, but time is often a big problem and urgent solutions must be found.

6. Prospects for new technologies

It is unfortunate that at the beginning of the 2020's it is still not possible to recommend a method to achieve the so desirable economic and environmental conversion of biomass into ethanol. There is an urgency to find a good method, even now while writing these words, because alcohol is considered an essential good worldwide as a disinfectant in the fight against the spread of the SARS-CoV-2 Virus (COVID-19). Much research effort has been made in the last 20 years with not very significant advances, which seems to point to the need to search for new directions.

This book and its authors thought that a different concept of biorefinery could be more effective. The thermal conversion of biomass is another possibility that can overcome the difficulties encountered in the use of biomass and to which other waste, such as human and garbage may be added.

The intricate and strong structure of biomass can be dismantled by producing materials that are also useful, but different from ethanol, at the same time producing sugars as a by-product capable of being fermented in ethanol. The disassembly of biomass can be done gradually and in order to take advantage of the properties of each fraction– most as new biomaterials with a guaranteed market and in the growth phase—producing some bioproducts and eventually less ethanol. Resins, waxes, and lipids are very easy to extract and, in its presence, they must be removed first. Hemicellulose is easy to extract at levels of 90% and new applications for its use appear every day (Xinxin et al. 2019). The remaining lignocellulosic material is cellulose and lignin; the latter can be used in formulations of polymer and carbon fiber or other resins; in its hydrolyzed form it acts as an effective and inexpensive antioxidant. The remaining cellulose can yield microcrystalline cellulose, MCC and, in advance, CNC nanocellulose particles or fibers (CNF), at the same time, releasing glucose capable of being fermented in ethanol. This is a possible scheme among many others. This relevant information is provided in the subsequent chapters.

7. References

Adebayo, E. A. and D. Martinez-Carrera. 2014. Oyster mushrooms (Pleurotus) are useful for utilizing lignocellulosic biomass. African Journal of Biotechnology 14: 52–67.

Alves, S. L. Jr., Herberts, R. A., Hollatz, C., Trichez, D., Miletti, L. C., de Araujo, P. S. and B. U. Stambuk. 2008. Molecular analysis of maltotriose active transport and fermentation by saccharomyces cerevisiae reveals a determinant role for the AGT1 permease. Applied and Environmental Microbiology 74(5): 1494–501.

Ankit, S., Shashi, P., Sunder, D., Sneh, G., Kirti, S., Seigo, A. and I. Kazuyuki. 2012. Bioethanol production from xylose: Problems and possibilities. Journal of Biofuels 3: 1–17.

Antunes, F. A. F., Santos, J. C., Chandel, A. K., Carrier, D. J., Peres, G. F. D., Milessi, T. S. S. and S. S. da Silva. 2019. Repeated batches as a feasible industrial process for hemicellulosic ethanol production from sugarcane bagasse by using immobilized yeast cells. Cellulose: 3787–3800.

Anwarab, Z., Gulfraz, M. and M. Irshada. 2014. Agro-industrial lignocellulosic biomass a key to unlock the future bio-energy: A brief review. Journal of Radiation Research and Applied Sciences 7: 163–173.

Arai, T., Biely, P., Uhliarikova, I., Sato, N., Makishima, S., Mizuno, M., Kaneko, S. and Y. Amano. 2019. Structural characterization of hemicellulose released from corn cob in continuous flow type hydrothermal reactor. Journal of Bioscience and Bioengineering 127: 222–230.

Baath, J. A., Mazurkewich, S., Poulsen, J. N., Olsson, L., Lo Leggio, L. and J. Larsbrink. 2019. Structure-function analyses reveal that a glucuronoyl esterase from Teredinibacter turnerae interacts with carbohydrates and aromatic compounds. Journal of Biological Chemistry 294: 6635–6644.

Bekedam, E. K. 2008. Coffe Brew Melanoidins. PhD Thesis, Wageningen University. The Netherlands.

Boeriu, C. G., Bravo, D., Gosselink, R. J. A. and J. E. G. van Dam. 2004. Characterisation of structure-dependent functional properties of lignin with infrared spectroscopy. Industrial Crops and Products 20: 205–218.

Bottger, C. and K. H. Sudekum. 2018. Review: protein value of distillers dried grains with solubles (DDGS) in animal nutrition as affected by the ethanol production process. Animal Feed Science and Technology 244: 11–17.

Bregado, J. L., Secchi, A. R., Tavares, F. W., Rodrigues, D. D. and R. Gambetta. 2019. Amorphous paracrystalline structures from native crystalline cellulose: A molecular dynamics protocol. Fluid Phase Equilibria 491: 56–76.

Caliari, Í. P., Márcio, H. P., Barbosa, M. H., Ferreira, S. O. and R. F. Teófilo. 2016. Estimation of cellulose crystallinity of sugarcane biomass using near infrared spectroscopy and multivariate analysis methods. Carbohydrate Polymers 158: 20–28.

Deswal, D., Gupta, R., Nandal, P. P. and R. C. Kuhad. 2014. Fungal pretreatment improves amenability of lignocellulosic material for its saccharification to sugars. Carbohydrate Polymers 99: 264–269.

Ding, D. Y., Zhou, X., You, T. T., Zhang, X., Zhang, X. M. and F. Xu. 2018. Exploring the mechanism of high degree of delignification inhibits cellulose conversion efficiency. Carbohydrate Polymers 181: 931–938.

Erdei, B., Franko, B., Galbe, M. and G. Zachi. 2012. Separate hydrolysis and co-fermentation for improved xylose utilization in integrated ethanol production from wheat meal and wheat straw. Biotechnology for Biofuels 5: 12.

Fink, H. P., Philipp, B., Paul, D., Serimaa, R. and T. Paakkari. 1987. The Structure of amorphous cellulose as revealed by wide-angle X-ray-scattering. Polymer 28: 1265–1270.

Furquim Da Cruz, A. 2013. Mannan-degrading enzyme system. *In*: Maria de Lourdes, T. M. and Polizeli, Mahendra Rai (eds.). Fungal Enzymes, CRC Press, Boca Raton, Fl.

Geng, W., Narron, R., Jiang, X., Pawlak, J. J., Chang, H., Park, S., Jameel, H. and R. A. Venditti. 2019. The influence of lignin content and structure on hemicellulose alkaline extraction for non-wood and hardwood lignocellulosic biomass. Cellulose 26: 3219–3230.

Ghanghas, R., Jindal, A. and S. Vasudevan. 2020. Geometry of hydrogen bonds in liquid ethanol probed by proton NMR experiments. Journal of Physical Chemistry B 124: 662–667.

Goldberg, R. N., Schliesser, J., Mittal, A., Decker, S. R., Santos, A. F. L. O. M., Freitas, V. L. S., Urbas, A., Lang, B. E., Heiss, C., da Silva, M. D. M. C. R., Woodfield, B. F., Katahira, R., Wang, W. and D. K. Johnson. 2015. A thermodynamic investigation of the cellulose allomorphs: Cellulose (am), cellulose Iβ(cr), cellulose II(cr), and cellulose III(cr). Journal of Chemical Thermodynamics 81: 184–226.

Guo, J., Huang, K., Zhang, S. and X. Yong. 2020. Optimization of selective acidolysis pretreatment for the valorization of wheat straw by a combined chemical and enzymatic process. Journal of Chemical Technology and Biotechnology 95: 694–701.

Hallac, B. B. and A. J. Ragauskas. 2011. Analyzing cellulose degree of polymerization and its relevancy to cellulosic ethanol. Biofuels Bioprod. Bioref. 5: 215–225.

Harris, D. and S. DeBolt. 2008. Relative crystallinity of plant biomass: Studies on assembly, adaptation and acclimation. Plos One 3: e2897.

Huang, S., Kiemle, S. N., Makarem, M. and S. H. Kim. 2020. Correlation between crystalline cellulose structure and cellulose synthase complex shape: a spectroscopic study with unicellular freshwater alga micrasterias. Cellulose 27: 57–69.

Inokuma, K., Iwamoto, R., Bamba, T., Hanasunuma, T. and A. kondo. 2017. Improvement of xylose fermentation ability under heat and acid co-stress in saccharomyces cerevisiae using genome shuffling technique. Frontiers in Bioengineering and Biotechnology 5: 81.

Isikgor, F. H. and C. R. Becer. 2015. Lignocellulosic biomass: a sustainable platform for the production of bio-based chemicals and polymers. Polym. Chem. 6: 4497–4559.

Jouzani, G. S. and M. J. Taherzadeh. 2015. Advances in consolidated bioprocessing systems for bioethanol and butanol production from biomass: a comprehensive review. Biofuel Research Journal-BRJ 2: 152–195.

Koizumi, S., Yue, Z., Tomita, Y., Kondo, T., Iwase, H., Yamaguchi, D. and T. Hashimoto. 2008. Bacterium organizes hierarchical amorphous structure in microbial cellulose. European Physical Journal E 26: 137–142.

Kontturi, E. 2015. Cellulose nanocrystals: New preparation routes, and the relationship to the structure of native cellulose. Abstracts of Papers of the American Chemical Society 249: 4. 249th ACS National Meeting. Denver, USA.

Kumara, B. and A. V. Behera. 2018. Bioenergy for Sustainability and Security, Springer, New York. USA.

Kundu, P., Manna, B., Majumder, S. and A. Ghosh. 2019. Species-wide metabolic interaction network for understanding natural lignocellulose digestion in termite gut microbiota. Scientific Reports 9: 16329.

Lema-Ruminska, J., Kulus, D., Tymoszuk, A., Varejao, J. M. T. B. and K. Bahcevandziev. 2019. Profile of secondary metabolites and genetic stability analysis in new lines of Echinacea purpurea (L.) Moench micropropagated via somatic embryogenesis. Industrial Crops and Products 142: 111851.

Li, F., Xie, G., Huang, J., Zhang, R., Li, Y., Zhang, M., Wang, Y., Li, A., Li, X., Xia, T., Qu, C., Hu, F., Ragauskas, A. J. and L. Peng. 2017. OsCESA9 conserved-site mutation leads to largely enhanced plant lodging resistance and biomass enzymatic saccharification by reducing cellulose DP and crystallinity in rice. Plant Biotechnol. J. 15: 1093–1104.

Li, M., Pu, Y. Q. and A. J. Ragauskas. 2016. Current understanding of the correlation of lignin structure with biomass recalcitrance. Frontiers in Chemistry 4: 45.

Li, M., He, B., Li, J. and L. Zhao. 2019. Microcrystalline cellulose sources. Bioresources 14: 7886–7900.

Linder, A., Bergman, R., Bodin, A. and P. Gatenholm. 2003. Mechanism of assembly of xylan onto cellulose surfaces. Langmuir 19: 5072–5077.

Lommerse, J. P. M., Price, S. L. and R. Taylor. 1997. Hydrogen bonding of carbonyl, ether, and ester oxygen atoms with alkanol hydroxyl groups. Journal of Computational Chemistry 18: 757–774.

Luo, Y. P., Li, Z., Li, X. L., Liu, X. F., Fan, J. J., Clark, J. H. and C. W. Hu. 2019. The production of furfural directly from hemicellulose in lignocellulosic biomass: A review. Catalysis Today 319: 14–24.

Lynd, L. R., Liang, X. and M. J. Biddy. 2017. Cellulosic ethanol: status and innovation. Current Opinion in Biotechnology 45: 202–211.

Mahmood, Z., Yameen, M., Jahangeer, M., Riaz, M., Ghaffar, A. and I. Javid. 2018. Lignin as Natural Antioxidant Capacity. In Lignin - Trends and Applications. InTech.

Meents, M. J., Motani, S., Mansfield, S. D. and A. L. Samuels. 2019. Organization of xylan production in the golgi during secondary cell wall biosynthesis. Plant Physiology 181: 527–546.

Mittal, A., Katahira, R., Himmel, M. E. and David K. Johnson. 2011. Effects of alkaline or liquid-ammonia treatment on crystalline cellulose: Changes in crystalline structure and effects on enzymatic digestibility. Biotechnol. Biofuels 4: 41.

Miura, K. and T. Nakano. 2015. Analysis of mercerization process based on the intensity change of deconvoluted resonances of C-13 CP/MAS NMR: Cellulose mercerized under cooling and non-cooling conditions. Materials Science & Engineering C-Materials for Biological Applications 53: 189–195.

Mohammad, A. S., Biernack, J., Scott, N. and M. Adenson. 2018. Diffusion of CO_2 and fractional free volume in crystalline and amorphous cellulose. Journal of Analytical and Applied Pyrolysis 134: 43–51.

Naidu, D. S., Hlangothib, S. P. and M. J. Johna. 2018. Bio-based products from xylan: A review. Carbohydrate Polymers 179: 28–41.

Nelson, M. L. and R. T. O'Connor. 1964. Relation of certain infrared bands to cellulose crystallinity and crystal latticed type. Part I. Spectra of lattice types I, II, III and of amorphous. Cellulose 8: 1311–1324.

Nomura, S., Sato, S. and T. Erata. 2020. DFT approach to the pathway of conformational changes of cellulose C6-hydroxymethyl group with simple cellotetraose model involving the mechanism of mercerization process. Chemical Physics Letters 742: 137154.

Ottesen, V., Larsson, P. T., Chinga-Carrasco, G., Syverud, K. and O. W. Gregersen. 2019. Mechanical properties of cellulose nanofibril films: effects of crystallinity and its modification by treatment with liquid anhydrous ammonia. Cellulose 26: 6615–6627.

Peng, X., Junli, R. and L. Zhong. 2011. Homogeneous synthesis of hemicellulosic succinates with high degree of substitution in ionic liquid. Carbohydrate Polymers 86: 1768–1774.

Puls, J. 1997. Chemistry and biochemistry of hemicelluloses: Relationship between hemicellulose structure and enzymes required for hydrolysis. Macromolecular Symposia 120: 183–196.

Ralph, J., Lapierre, C. and W. Boerjan. 2019. Lignin structure and its engineering. Current Opinion in Biotechnology 56: 240–249.

Rastogi, M. and S. Shrivastava. 2018. Recent advances in second generation bioethanol production: An insight to pretreatment, saccharification and fermentation processes. Journal of Biotecnhology & Research 1: 1–8.

Rose, J. K. C. 2003. Plant Cell Wall: Biosynthesis, Structure and Function. Ed., Pub. Blackwell Publishing. CRC Press. Boca Raton, USA.

Saini, J. K., Patel, A. K. and M. Adsul. 2015a. Cellulase adsorption on lignin: A roadblock for economic hydrolysis of biomass. Renewable Energy 98: 29–42.

Saini, J. K., Saini, R. and L. Tewari. 2015b. Lignocellulosic agriculture wastes as biomass feedstocks for second-generation bioethanol production: concepts and recent developments. Biotech. 5: 337–353.

Simmons, T. J., Mortimer, J. C., Oigres, W., Bernardinelli, D., Poppler, A. C., Brown, S. P., Azevedo, E. R., Dupreel, R. and P. Dupree. 2016. Folding of xylan onto cellulose fibrils in plant cell walls revealed by solid-state NMR. Nature Communications 7: 13902.

Singla, A., Paroda, S., Dhamija, S. S., Goyal, S., Shekhawat, K., Amachi, S. and K. Inubushi. 2012. Bioethanol production from xylose: Problems and possibilities. Journal of Biofuels 3: 39–49.

Song, Y., Zhang, J., Zhang, X. and T. Tan. 2015. The correlation between cellulose allomorphs (I and II) and conversion after removal of hemicellulose and lignin of lignocellulose. Bioresource Technology 193: 164–170.

Suhas, Carrott P. J. M. and M. M. L. Ribeiro Carrott. 2007. Lignin—from natural adsorbent to activated carbon: A review. Bioresour. Technol. 98: 2301–12.

Sung, Y. J. and J. Shin Soo. 2011. Compositional changes in industrial hemp biomass (Cannabis sativa L.) induced by electron beam irradiation pretreatment. Biomass & Bioenergy 35: 3267–3270.

Talebi Amiri, M., Bertella, S., Questell-Santiago, Y. M. and J. S. Luterbacher. 2019. Establishing lignin structure-upgradeability relationships using quantitative 1H–13C heteronuclear single quantum coherence nuclear magnetic resonance (HSQC-NMR) spectroscopy. Chem. Sci. 10: 8135.

van Soest, Traxler M. J. and D. G. Fox. 1998. Predicting forage indigestible NDF from lignin concentration. Journal of Animal Science 76: 1469–1480.

Varejão, J. M. T. B. 2001. Towards the synthesis of model peroxisases. PhD Thesis, University of Liverpool, UK.

Varejão, J. M. T. B. and Nazaré. 2017. Ethanol production from macroalgae biomass. In: Leonel Pereira (ed.). Algal Biofuels. CRC Academic Press. Boca Raton. USA.

Vilela, L. F., Gomes de Araujo, V. P., Paredes, R. S., Bon, E. P., Torres, F. A., Neves, B. C. and E. C. Eleutherio. 2015. Enhanced xylose fermentation and ethanol production by engineered Saccharomyces cerevisiae strain. Amb Express 5: 16.

Wada, M., Ike, M. and K. Tokuyasu. 2010. Enzymatic hydrolysis of cellulose I is greatly accelerated via its conversion to the cellulose II hydrate form. Polym. Degrad. Stab. 95: 543–548.

Wang, T. and M. Hong. 2016. Solid-state NMR investigations of cellulose structure and interactions with matrix polysaccharides in plant primary cell walls. Journal of Experimental Botany 67: 503–514.

White, P. M., Viator, R. and C. L. Webber. 2020. Temporal and varietal variation in sugarcane post-harvest residue biomass yields and chemical composition. Industrial Crops and Products 154: 112616.

Wu, X., Wagner, R., Raman, A. and M. A. Moon. 2010. Elastic Deformation Mechanics of Cellulose Nanocrystals. TMS 2010 Annual Meeting Supplemental Proceedings on Materials Processing and Properties. 139th annual meeting & exhibition—supplemental proceedings, vol 2: materials characterization, computation and modeling and energy. Seattle, WA.

Wu, Z., Xu, J., Li Gong, J. and L. Mo. 2018. Preparation, characterization and acetylation of cellulose nanocrystals allomorphs. Cellulose 25: 4905–4918.

Xiao, B., Sun, X. F. and S. R. un Cang. 2001. Chemical, structural, and thermal characterizations of alkali-soluble lignins and hemicelluloses, and cellulose from maize stems, rye straw, and rice straw. Polymer Degradation and Stability 74: 307–319.

Xinxin, L. Q., Lina, Y., Yana, Y., Pengb, F., Sunb, R. and J. Rena. 2019. Hemicellulose from plant biomass in medical and pharmaceutical application: A critical review. Current Medicinal Chemistry 26: 2430–2455.

Yao, L., Yong-Chao, L., Hong-Qin, H., Feng-Jin, X., Xian-Yong, W. and F. Xing. 2017. Structural characterization of lignin and its degradation products with spectroscopic methods. Spectroscopy in Fuels 2017: 8951658.

Yong, T. L. K., Khalid, K. A. and A. A. Ahmad. 2017. Lignin extraction from lignocellulosic biomass using sub- and supercritical fluid technology as precursor for carbon fiber production. Journal of Japan Institute of Energy 96: 255–260.

Yu, Z. H. and W. Klemperer. 2005. Asymmetry in angular rigidity of hydrogen-bonded complexes. Proceedings of the National Academy of Sciences of the United States of America 102: 12667–12669.

Zhang, G. C., Liu, J. J., Kong, I. I., Kwak, S. and Y. S. Jin. 2015a. Combining C6 and C5 sugar metabolism for enhancing microbial bioconversion. Current Opinion Chemical Biology 29: 49–57.

Zhang, J., Choi, Y. S., Yoo, C. G., Kim, T. H., Brown, R. C. and B. H. Shanks. 2015b. Cellulose–hemicellulose and cellulose–lignin interactions during fast pyrolysis. ACS Sustainable Chemistry & Engineering 3: 293–301.

Zhao, L., Zhang, X. and J. Xu. 2015. Techno-economic analysis of bioethanol production from lignocellulosic biomass in china: dilute-acid pretreatment and enzymatic hydrolysis of corn stover. Energies 8: 4096–4117.

Chapter 2

New Uses for Hemicellulose

1. Introduction

Hemicellulose constitutes approximately 30–45% of the biomass dry weight. Due to its limited connections to cellulose and lignin, essentially of a hydrogen bonding nature, it is one in three main components of biomass that is relatively easier to remove. Solvents that can form hydrogen bonds such as water, alcohols and ammonia are suitable for their extraction. The oldest extraction methods used acid solutions that lead to the hydrolytic extraction of hemicellulose, useful in cases where the objective is the recovery of sugars. Extraction in non-hydrolytic conditions, obtaining hemicellulose in its native form, only began to be studied in the last decade. In the literature, especially from 2015 onwards, studies on the use of hemicellulose in the medical and pharmaceutical areas began to appear, in which hemicellulose is used in its non-hydrolyzed form (Liu et al. 2016).

The subgroup of hemicellulose with the highest natural occurrence in the biomass is xylan, in this, the most abundant monosaccharide is the pentose (C5) xylose, with the presence in smaller amounts of arabinose, glucose and mannose, among others. Xylan is found mainly in hardwood and grass biomass. In some tissues of specific grass species, the xylose content can represent up to 50% of the carbohydrate fraction (Kuhad et al. 2011). The soft wood and the group of resinous plants present another type of hemicellulose—the subtype of glucomannans (Du et al. 2014).

2. Extraction of hemicellulose from biomass in pristine form

Several new emerging applications in which hemicellulose could play an essential role can be observed only when it can be extracted from the cell wall in its native form, that is, not hydrolyzed and with a degree of polymerization (DP) close to the original. For this to happen, labile conditions extraction method should be provided. The position of hemicellulose in the biomass is in the domain it forms with the lignin, on rare occassions it is close to and interacts with the cellulose fibers. The binding of hemicellulose both to cellulose and lignin occurs mainly through hydrogen bonding at the interface between the three components. Subordinately, covalent bonds at certain points with lignin have been referred to. This means that the extraction of hemicellulose is possible, in different types of biomass, with the use of solvents capable of forming hydrogen bonds, the most common being aqueous ammonia, alcohols or even hot water. These hydrothermal methods show that partial recovery of hemicellulose can be done in the absence of any acid to avoid breaking the glycosidic bond forming the backbone of the pentose based polymer. The temperature in these hydrothermal extractions must not exceed temperatures of 150–160°C otherwise several types of reactions can occur, namely the glycosidic bond hydrolysis with an increasing formation of xylooligomers and simple sugars, the last which under these conditions undergo rapid degradation (Gallina et al. 2016). Treatment time is another important variable, prolonged extractions (> 1 h) should be avoided, as they usually lead to an increase in the degree of hydrolysis and a reduction in the molecular weight of native hemicellulose. Hemicellulose has the $\beta(1,4)$ bond, which in the case of pentoses is relatively easy to hydrolyze. This recommends caution in hydrothermal extraction, with an in-depth study of the conditions for each type of biomass. Even when using only pure water in the hydrothermal treatment, it tends to become acidic due to the presence of multiple acetyl groups (CH_3CO_2-) in the hemicellulose, which by hydrolysis make the medium acidic at concentrations that favor the hydrolysis of the polymer. To avoid this effect, the use of slightly basic conditions is recommended. Yields tend to be low, in the range of 35–60% when extracting hemicellulose under labile conditions, its increase requires changes in the extraction process.

The arrangement of hemicellulose in the general structure of the biomass presents in some points fibers intertwined with lignin, with probable covalent bonds, turning its complete extraction difficult. Studies showed that there is a negative correlation between the higher relative content of lignin and the yield of hemicellulose extraction (Geng et al. 2019). If before the hydrothermal extraction, the hydrolysis of the hemicellulose-lignin bonds could be done, a higher yield can be obtained. To prevent damage to the structure of hemicellulose, a preliminary treatment specifically directed to the partial degradation of lignin can be done with the use of enzymes, such as manganese peroxidases or their commonly available synthetic versions (Rao et al. 2019). The use of acid may constitute another alternative, specially if made in short periods of time. The use of organic weak acid may to a degree, facilitate the hydrolysis of hemicellulose with some covalent bound to lignin and improve removal efficiency, but with limitations, since the structure of hemicellulose may be affected. Breaking these leads to changes in the hemicellulose skeleton, which in turn could origin undesirable loss of structure. In fact, references have been made that the properties of hemicellulose obtained from biomass are dependent on the extraction method used (Sannigrahi et al. 2010, Naidu et al. 2018).

Fortunately, hydrolysis with weak base solutions can extract hemicellulose under satisfactory conditions with a low degree of structural modification. With its use, extraction yields can reach values of 80–90% (Naidu et al. 2018). An example uses slightly alkaline NaOH solutions, capable of neutralizing the release of acetic acid, at the same time carrying out the alkaline hydrolysis of hemicellulose-lignin bonds.

Depending on the extraction method used the final hemicellulose extract may show characteristics from of almost intact hemicellulose polymeric material to a mixture of pristine hemicellulose with some content of hydrolyzed moieties, as xylooligosaccharides (XOS) in nature.

A compilation of experimental results in the extraction of hemicellulose from biomass, in an almost pristine form, is shown in Table 1.

Table 1. Examples of "intact" hemicellulose extraction experiments on different types of biomass.

Biomass type	Extraction method	Hemicellulose yield (% dry mass)	Notes	Reference
Different woods	Recirculating hot water, 160°C	44	10–60 kD polymers	Gallina et al. 2016
Wheat straw	10% aqueous NaOH, 100°C	80	Solid/liquid mass ratio 20:1	Geng et al. 2019
Sugarcane waste	10% aqueous NaOH, 100°C	90	Solid/liquid mass ratio 20:1	Geng et al. 2019
Spruce and Birch wood	Hot water, 160°C	68–75		Giummarella et al. 2017
Tobacco biomass	Hot water, 200°C, microwave	10	13 to 144 kDa	Yuan et al. 2019

2.1 Medical and nutritional uses

Xylooligosaccharides (XOS) cannot be used by the microorganisms *Streptococcus mutans* that are usually found in the human mouth and that can give rise to the formation of dental caries. XOS are also known to inhibit the activity of glucose transferase, a key enzyme in the production of glucan polymers, substances involved in the formation of tartar, mediated with extracellular protein from *S. mutans* (Lynch et al. 2013). Effects such as preventing tooth decay and fresh breath have also been associated with the incorporation of xylooligosaccharides in toothpaste (Seo et al. 2004).

To hemicellulose, strong adsorbent capabilities of cholesterol and bile acids have also been attributed, so that its oral administration can reduce the cholesterol content in the blood (Hu and Yu 2013).

Xylooligosaccharides can selectively favor the proliferation of Bifidobacterium bacteria in the intestine, to which factors that enhance the phagocytic activity of lymphocytes are attributed, allowing for the direct elimination of tumor cells. The branching of the hemicellulose polysaccharide structure blocks its enzymatic degradation by the glucosidases present in the human stomach, making hemicellulose of low digestibility. Human nutrition studies show that XOS pass largely untouched through the small intestine into the large intestine. Several variants of Bifidobacterium probiotics and lactic acid bacteria can metabolize hemicellulose, together with the fermentation of short-chain fatty acids, such as acetic, propionic and butyric acid, leading to a reduction in the pH value of the intestine, an effect that is related to good health intestinal function, with benefits in lipid metabolism, mineral absorption and lower prevalence of colon cancer (Koropatkin et al. 2015). The inhibition of the growth of *Escherichia coli*, *Clostridium* and other pathogenic bacteria has been demonstrated by the use of XOS, playing a positive role in adjusting the balance of intestinal flora (Le et al. 2020).

The aforementioned difficult enzymatic attack on hemicellulose polymers in the human digestion process, makes them stable to gastric conditions, making it possible for them to pass into the intestine. This makes it possible to manufacture biocompatible polymers, to release specific drugs in the intestine, by oral ingestion (Sauraj et al. 2017).

Hemicellulose polymers can serve as a raw material for preparing hydrogels for generic medical uses (Fonseca et al. 2011, Svard et al. 2015). In Figure 1, a summary of the main research routes for the use of hemicellulose is presented.

Hemicellulose is also important in animal feed, due to its greater digestibility in relation to cellulose, both in ruminants and monogastric animals. It is common sense that herbivorous animals prefer grass and straw from different cereals, as they are rich in hemicellulose, as a food source. Pig production can also benefit from the addition of nutritional supplements rich in hemicellulose (Tiwari et al. 2019).

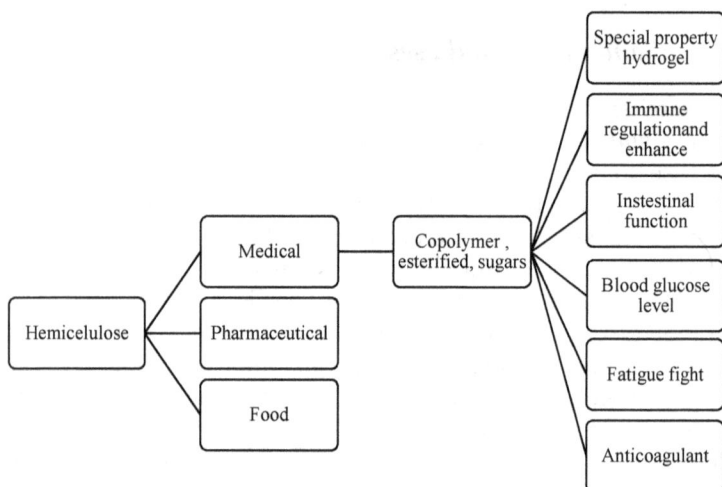

Figure 1. List of ongoing trials for the development of the use of unmodified hemicellulose fraction from residual biomass. Adapted from Liu et al. (2019).

2.2 *Other uses*

Hemicellulose has been used as an adsorbent material capable of metals extraction in the purification of contaminated water. For this, variants that possess some extent of glucuronic acid and deacetylation by chemical or enzymatic methods must be performed; free acetate groups reinforce hemicellulose chelating properties, making it suitable for removing heavy metals in a slightly alkaline medium, a process that is parallel to the use of chitin for the same purpose (Dax et al. 2015).

3. Hydrolytic extraction of hemicellulose

The use of acids or bases in more severe conditions, meaning higher concentration and temperature, leads to the hydrolytic extraction of hemicellulose. In this method, hemicellulose's native structure is broken by cuts at different points giving rise to a complex mixture of small oligosaccharides/monosaccharides that diffuse from biomass. This process is suitable for uses where the objective is the saccharification of hemicellulose to obtain essentially the sugars or products of their degradation. The eventual fermentation of sugars to obtain ethanol, the preparation of xylitol or a production of hydroxymethylfurfural (HMF) are examples in which this method serves the purpose.

Various acids are efficient in saccharification, with diluted sulfuric or phosphoric acid being particularly effective. The operational concentration range starts at concentrations of 5 to 10%, up to 15–17%, in aqueous solution, at temperatures of 100°C to 250°C, in the latter case in a pressurized vessel. Other useful techniques are the impregnation of acids with simultaneous mechanical action, followed by aqueous extraction of sugars. In the latter, fine control of experimental conditions is necessary, especially with the amount of acid added, as its excess can destroy the pentoses. Heating using microwave radiation, in the presence of acid, can be particularly useful due to its catalytic hydrolytic properties (Chadni et al. 2019). With these conditions, it is possible to obtain sugar contents with yields of 80–90% based on the initial hemicellulose content. Xylan, as an example of a pentose—in which the sixth carbon is absent—a characteristic that facilitates the hydrolysis of the $\beta(1,4)$ glycosidic bond relative to cellulose. Traces of furfural can be produced. If fermentation is thought of as the subsequent process, this can prove to be difficult to perform, and the furfural content must be evaluated and removed. A slight contamination of the hydrolysates with acetic acid, phenols and phenolic acids is expected, considering their presence in the hemicellulose/lignin fractions.

As the acid and temperature conditions rise, other biomass fractions start to hydrolyze, as occur with the cellulose amorphous zones, but more importantly, already formed pentoses begin to degrade, which have little acid stability; producing furfural and subsequently other condensation substances derived from aldehydes and even substances related to coals. The high reactivity of C5 sugars with acids—for example sulfuric acid— can even carbonize the hemicellulose fraction in a relatively easily fashion.

Figure 2 shows an HPLC chromatogram of a hydrolyzate from a dry (60°C) and ground (1 mm), sample of corn stalk using 5% aqueous sulfuric acid, for 1 h, at 100°C with microwave heating. The sugar bands are expressed as a relative percentage, with their chromatographic retention time in parentheses, and are xylose (81%), arabinose (4%), mannose (5%) and glucose (10%).

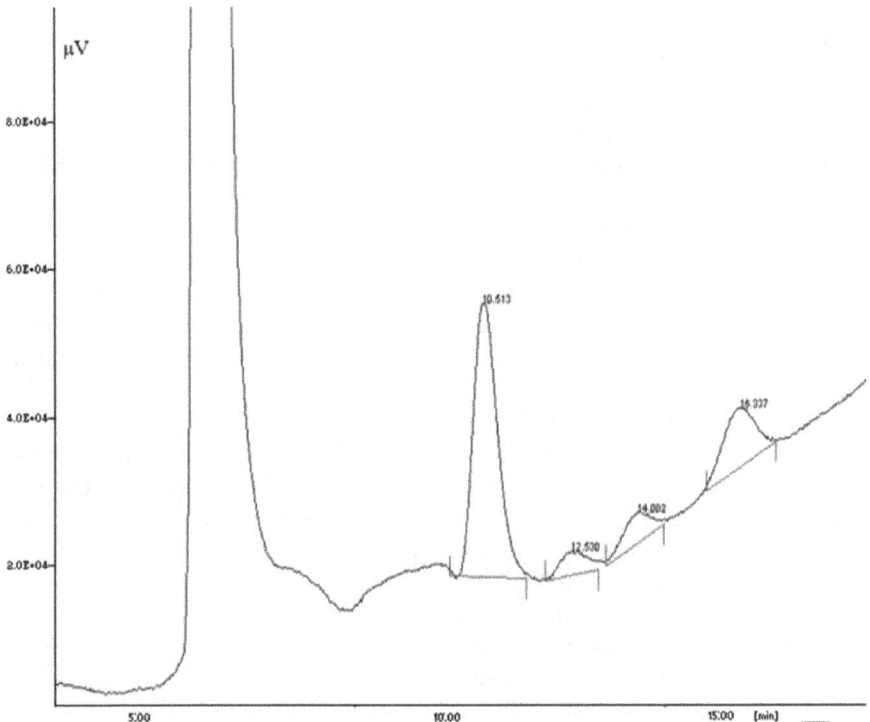

Figure 2. HPLC chromatogram of corn stem hydrolyzate, using 5% sulfuric acid, for 1 h, under microwave irradiation at 100°C. The chromatographic bands are, xylose (retention time, tr = 10.51 min); arabinose (tr = 12.53 min); mannose (tr = 14.00 min); glucose plus galactose (retention time, tr = 16.37 min).

4. Fermentation of hemicellulose to ethanol

The great difficulty in fermenting the pentoses in ethanol has been one of the reasons for the delay in pursuing the highly desirable cellulosic ethanol prepared from different types of biomass residues. Hemicellulose is responsible for 30% of the dry matter of the biomass, which cannot be wasted in the fermentation of the biomass for a competitive conversion into ethanol. In nature, there are microorganisms capable of using C5 sugars as a raw material to obtain metabolic energy, releasing ethanol as a by-product. The metabolic pathway of xylose in bacteria involves the isomerization of xylose into xylulose by the enzyme xylulose isomerase with subsequent phosphorylation. The latter species can enter the metabolism of the pentose phosphate pathway (PPP), producing ethanol (Maurya et al. 2015). Yeasts native species most capable to date of carrying out xylose fermentation are *Candida shehatae* (Dall Cortivo et al. 2020), *Pachysolen tannophilus* (da Silva et al. 2020) and *Pichia stipites* (Jeong et al. 2012). In these, the metabolism pathway involves a first reduction of xylose in xylitol, promoted by the enzyme xylitol reductase, followed by conversion into xylulose by the enzyme xylitol dehydrogenase. The major problem with the yeasts mentioned above is their low tolerance to ethanol, which at best does not exceed concentrations of 6 to 7% of ethanol by volume. In this respect, several strains of *S. cerevisiae* are viable in concentrations of ethanol up to about 14% (v/v) facilitating the ethanol separation from aqueous solution. If xylose could be transformed into xylulose in the yeast *S. cerevisiae*, this sugar would be possible to be introduced into its glycolytic pathway, benefiting from its excellent characteristics for ethanol production. The genes that encode the two enzymes responsible for xylose metabolism do not exist in the *S. Cerevisiae* genome. In the last decades, a great effort has been made to insert the genes that encode those enzymes into its genome, mainly in plasmids. The objective was successful, but the modified microorganisms showed a low production of ethanol. Cross interference between enzymes at the level of inhibition in the metabolic pathways occurs. New variants have been developed, but they are not able to consume xylose until the glucose is completely depleted (Vilela et al. 2015).

However, improvements have been achieved with other strategies, such as improving the tolerance of strains to higher values of temperature and concentration of organic acids in the fermentation medium. These new strains allow simultaneous saccharification and co-fermentation (SSF process, see below) (Inokuma et al. 2017). The search for new microorganisms and overcoming the difficulties now encountered will eventually happen with a greater research effort. The thermotolerant yeast *Spathaspora passalidarum* CMUWF1-2 (associated with the flat wood beetle) is an example of a microorganism that is not repressed by the presence of glucose and may co-ferment sugars in C5 and C6 (Rodrussamee et al. 2018).

5. Adding a carbon to pentose sugars

Monosaccharides, such as xylose or arabinose, are aldehydes—organic compounds known for their strong reducing properties and reactivity. Adding groups to the carbonyl group of the aldehyde is a well-known branch of organic chemistry. If ethanol production is the goal from C5 sugars, a new possibility may emerge, combining chemistry with biochemistry, and may have the strength to potentially constitute a means of converting C5 sugars into ethanol. Several chemical reactions make possible to add a group,CH2, to the pentoses, converting them to hexoses, if necessary with another hydroxyl group –CH2OH. A problem arises in this addition of carbon—sugars and Nature exercise strict control over stereocenters in their building blocks— and chemistry often has difficulty making reactions in a stereo-specific manner. The insertion of one more carbon should ideally be done so that the final sugar—a hexose—results in the D configuration, that is only one of the two possible stereo geometries. In general, known reactions result in the formation of a racemic mixture. However, in the literature, there are cases in which microorganisms were able to convert stereoisomers of sugar-related substances into ethanol (Kamzolova et al. 2018) by expression of enzymes capable of rearranging their structures (Pikis et al. 2006). An example of this possibility occurs with arabinose, as shown in Figure 3, where the oxidative addition of a –CH2- group results in a mixture of D-Glucose plus D-Mannose. Since these sugars are interconvertible via fructose through the transformation of Lobry from Bruyn-van Ekenstein (Stahlberg et al. 2012), the fermentation of both to ethanol is possible. The addition of carbon can be done by different chemical reactions. The addition of sodium cyanide to aldose arabinose followed by reduction with hydrogen or aqueous acid in the presence of a suitable catalyst, produces a mixture of D-Glucose plus D-Mannose, via the Kiliani-Fischer reaction (Nishimura et al. 1994). The same products can be obtained using diisobutylaluminum hydride (DIBAL-H) route 2 at the top, Figure 3 (Ban et al. 2017). The use of bromodichloromethane may be another possibility by addition of the Grignard type to the carbonyl group and producing a formylation product by hydrolysis (Stepowska et al. 1999). A third possibility is shown in Figure 2 with the addition of carbon monoxide followed by reduction by catalytic hydrogenation (Chatani et al. 1995).

A drawback with the reductive addition of carbon is that when applied to the transformation of xylose, a mixture of D-Gulose and D-Idose is obtained. These sugars are not fermentable by common yeasts; however, they are considered to be rare sugars for which important properties in medical and human nutrition are thought to exist, although not fully studied

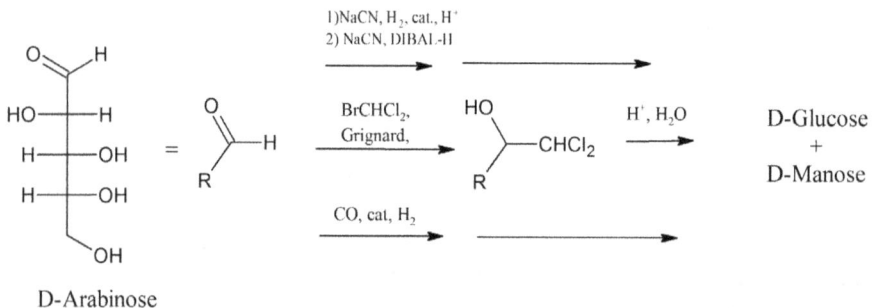

Figure 3. Examples of reaction pathways capable of transforming a pentose into a mixture of hexose stereoisomers.

(Chowdhury et al. 2015). In any case, they have the potential to be used in alternative areas with an economic value potentially greater than the final fermentation destination (Bilal et al. 2018).

6. Conversion of hemicellulose to xylitol

Xylose can be reduced to xylitol, a derivative that has been known for a long time and that has shown in recent years to be a valuable and desirable substance in different fields. The name xylose is derived from the Greek word xylon which means "wood"; xylose is also known as wood sugar, from which xylitol was first produced by reduction. First references to xylose have been around since the end of the 19th century, when in 1891 the German chemist Emil Fischer and his student Rudolf Stahel extracted a previously unknown substance from beech shavings, which was called "xylose" (Mäkinen 2000). From wheat and oat fibers, it is possible to isolate a syrup with sweet taste. For some years, xylose and its reduced form, xylitol, went unnoticed because it apparently had no scientific or commercial interest. During World War II, in Finland, there was a shortage of sugar and there was a need to look for substitutes. At that time, a natural sweetener was developed, using xylitol (Mäkinen 2000). Finnish engineers were able to start producing xylitol from biomass residues with a commercially viable manufacturing process, using cereal stalks as a raw material. Today, xylitol is prepared from various plant biomass waste by means of a relatively simple chemical or biotechnological process. The value and use of xylitol has generated a large number of research projects with the aim of improving its obtaining and spreading its use in new fields, thanks to results of scientific research on its biochemical properties.

Xylitol is considered an intermediate product of carbohydrate metabolism in humans, and found in blood in amounts of 0.03–0.06 mg/100 mL. It can also be found in nature in many fruits and vegetables, such as lettuce, cauliflower, plum, raspberry, strawberry, grape and banana, as well as in yeasts and mushrooms, in amounts less than 0.9 g/100 g (Rehman et al. 2013, Espinoza-Acosta 2020).

During the last decades of the previous century, consumption of xylitol grew steadily, with its area of use and market being expanded worldwide. It is currently approved for use as an additive in human foods, pharmaceuticals, oral health products (toothpaste, mouthwash) and other nutraceutical products. The world xylitol market reached a total value of around 1.4 million tons in 2013. The quantity used has grown continuously over the years, with a world production of around 242 thousand tons, and with an economic value above $ 1.3 billion in 2015.

6.1 Food/nutritional use

Xylitol is used as a sweetening and humectant food additive, that can be used in the amount necessary to obtain the desired effect, as long as it does not impair the genuineness and identity of the food. Xylitol has a sweetening power similar to that of sucrose, being 2.4 times sweeter than mannitol and 2.0 times more than sorbitol, a property that can be modulated by changing the pH value. It belongs to the type of substance known as polyols, with xylitol being responsible for the second largest portion of this market (about 12%), second only to sorbitol (48%). Xylitol, however, due to its unique properties, has the potential to grow at a higher rate than other polyols. Xylitol is an essential component in the preparation of products such as chewing gum, attracted by its cooling effect, a product that tends to grow with the increase in the world population. It is estimated to increase its use in confectionery and as a sweetener for people with diseases such as diabetes, and its consumption is growing at a high rate in recent years (Grembecka 2015).

Recently, legislation in most countries is limiting the amount of sugar added to drinks and juices or imposing heavy fees on their use. Obesity control has become an obligation in most countries due to changing population habits. This has led to a greater search for sweeteners in the food industry. Xylitol can replace sucrose, as it is low in calories and is safe for health. Xylitol has attracted the attention of beverage and food manufacturers, even though it is has a relatively high price. Thus, many products that incorporate this low calorie polyol are already available on the market, such as caramels, chocolates, etc. Other secondary uses are as a humectant, and as a food stabilizer, as it does not give the Maillard reaction, which is responsible for browning of food. It is also used as a cryoprotectant and freezing point reducer in food.

6.2 Use in personal care

Manufacturers have added xylitol to toothpaste, due to its cooling effect, derived from its positive dissolving enthalpy (+34.8 kcal/g), together with sweetening properties that are similar to sucrose. The cooling effect occurs on solubilization, requiring xylitol to be in solid form prior to its application in aqueous pastes (Gargouri et al. 2020).

Many studies over time have also proven that xylitol has a strong anticariogenic effect, as it is not metabolized by oral flora and gives the opportunity for more efficient dental remineralization (Espinoza-Acosta 2020). This sweetener together with its anti-carcinogenic action makes it a very desirable substance in the formulation of toothpaste and mouthwash elixir. Chewing gum is also recommended as a way of brushing teeth in work situations, in which oral hygiene after meals is sometimes not possible, in order to prevent the formation of cavities. The anticariogenic properties of xylitol are mainly determined by the non-fermentative action of bacteria of the genus Streptococcus, microorganisms that are not able to metabolize it, preventing its proliferation and, consequently, avoiding the production of organic acids that tend to corrode tooth enamel. As the concentration of *Streptococcus mutans* present decreases, there is a reduction in insoluble polysaccharides, favoring the increase in soluble polysaccharides, a means of making plaque less adherent and easily removed by the usual brushing of teeth (Gargouri et al. 2018). It can be highlighted that xylitol contributes to oral health in different ways: decreasing the incidence of cavities, stabilizing the content of calcium and phosphate ions in saliva, favoring dental remineralization; stabilization of already formed caries, reduction of the proliferation of *S. mutans* and *Lactobacillus* in saliva, favoring the formation of saliva (without increasing the production of acid in the dental plaque).

In the pharmaceutical industry, xylitol can be used as a sweetener or excipient in the formulation of syrups, tonics and vitamins. However, due to its high price, about 10 times that of sucrose or sorbitol, it is generally used in combination with other polyols that serve as mixing agents.

6.3 Medical uses

Xylitol is used in healthcare as a useful substance in areas such as the treatment of conditions such as diabetes, parenteral and kidney injuries, acute otitis media, lung diseases and osteoporosis (Salli et al. 2019). In general, long-term studies have associated health benefits in several fields with its use, making it a valuable substance. In the treatment of diabetes, xylitol, unlike conventional sugars, does not depend on insulin to be digested by the body, which leads to its good tolerance by people with diabetes. Thus, its concentration in the blood does not undergo sudden changes caused by sucrose and glucose, which makes xylitol a suitable sweetener for diabetics.

The effectiveness of xylitol in the treatment or prevention of osteoporosis has been reported in animal research, and it has been proven that xylitol favors the increase in bone mass, preserving the minerals contained in it, and preventing the weakening of bone mechanical properties (Mattila et al. 1998).

The use of xylitol as a food supplement in concentrations of 6 to 15% in some prepared foods has shown good results in the treatment of induced acute inflammatory processes. Although still in an earlier phase, research on the application of xylitol in curing or controlling inflammatory processes is very promising, since all indicate that, with a low content of xylitol in the diet, it is possible to obtain positive results in a short period of treatment, without impairing the general functioning of the organism (Park et al. 2014).

6.4 Xylitol obtention

The wide range of uses presented above makes obtaining xylitol a field of great research. With the increase in consumption follows demand and the increase in the world population requires that a greater quantity be produced annually. Xylitol is chemically a polyol, showing its molecule an acyclic structure containing four asymmetric carbons. The synthetic approach to substances with several stereochemical centers is always difficult to succeed, due to the production of racemic mixtures that require separations of enantiomers, the most difficult to find. Consequently, xylitol must be obtained from biological sources that possess it or that are rich in xylose. Biomass contains approximately 30% dry weight of hemicellulose, a polymer in which xylose is the predominant monomer. Thus, a viable source for obtaining xylose is biomass, especially if it can be collected from accumulated residues by existing processing, such as cereal straw or residues from biomass processing, such as cellulose pulp preparation. Its separation and isolation can be problematic, but more recent results have shown better possibility.

6.4.1 Obtaining xylose from biomass

Depending on local availability, various types of biomass residues can be used to recover xylose. The conditions that biomass must undergo for an efficient extraction of xylose are those for the extraction of hemicellulose under conditions that undergo hydrolysis, discussed above. The main method for removing hemicellulose is the use of diluted acid hydrolysis. Hydrochloric and sulfuric acids are the most used, although there are other acids that have been used as well. Since hemicellulose is amorphous, less severe conditions can be used to hydrolyze hemicellulose and recover monosaccharides with good specific yields.

More recently, studies have been conducted to use complex hemicellulolytic enzymes—xylanases—isolated from cultures of specific microorganisms. This is an alternative hydrolysis process that uses mild operating conditions, such as low temperature in the presence of buffer solutions, and which can potentially result in high yields of xylose, parallel to those obtained with the use of acids (You et al. 2019). However, there are problems in finding efficient processes in the recovery of xylose, requiring pre-treatment of biomass. These treatments aim to make it possible for enzymes to enter the structure of the hemicellulose/lignin domain of biomass. Examples of pre-treatments are hydrothermal processes, such as the use of water at high temperatures, explosion of steam or ammonia, and alkaline treatments that partially remove lignin from the cell wall leading to increased extraction/hydrolysis efficiency. This method has the advantage of not producing the yeast inhibitors characteristic of acid hydrolysis, namely furfural and 2-hydroxymethylfurfural (2-HMF). This is particularly relevant if the method chosen to reduce xylose to xylitol involves the use of yeast.

The main global suppliers of xylitol declared in 2010 (Danisco 2009, Anonymous 2010) that they were starting to produce xylitol chemically using xylose recovered from the sulphite liquors of the cellulose industry. These are the residue of the biomass treatment by NaOH/ sulfite at elevated temperatures, in the pulp industry. The announcement was published in 2010, when the company "informs that it will not continue to use biomass residues to obtain xylose, but to recover it from sulfite liqueurs, so that it does not imply with the final destination of the liqueurs that is its burning for supply energy and recovery of inorganic salts". The process is privately owned and many details about the process have never been revealed.

Taking into account the quantities of biomass that the cellulose industry processes worldwide per year, indicates that the cellulose industry is wasting and burning an amount of xylose that corresponds to about 20% of the processed biomass weight. By not converting it to xylitol, a substance of high economic value, and burning it, brings orth the rationale to develop complementary techniques for its isolation. Table 2 shows the chemical composition of the liqueurs of the sulfite liqueurs from the Kraft cellulosic process.

Table 2. Approximate chemical composition of thin (SSL) and thick (THSL) liquors of *Eucaliptus globulus* from a cellulose pulp production unit. Adapted from Esteves (Esteves et al. 2008).

Components	Sugars	% Liquor	
		SSL	THSL
Dry solids		12.8	56.8
ash		2.8	13.8
Furfural		0.2	traces
Acetic acid		0.8	0.3
Lignosulphonates		5.9	32.9
Total sugars		3.2	9.1
	Rhamose	0.1	0.2
	Arabinose	0.1	0.3
	Xylose	2.1	5.5
	Mannose	0.1	0.3
	Galactose	0.5	2.1
	Glucose	0.3	0.7

As can be seen, sugars represent 9% of concentrated liquors (THSL), with xylose being the main component ($\approx$ 62%). Looking at the chemical composition of the components of sulfite liqueurs, the vast majority, such as lignin sulfonates and inorganic salts have an ionic nature. Sugars are the only substances in the mixture not having an electric charge. In addition to the fact that there is little information in the literature on the separation of xylose from the sulphite liquors, the dialysis process or membrane separation specially designed using osmosis or reverse osmosis can be viewed as having the potential to do so, with minimal interference in the process already existing in the units. The wide use of these techniques will allow for the recovery of a greater amount of xylose and production of xylitol.

6.4.2 *Methods of reducing xylose to xylitol*

The production of xylitol from biomass with high levels of hemicellulose requires several steps. The appropriate biomass includes various types of straw, ears of corn and softwood, which contain high levels of xylan.

A detailed scheme for the preparation of xylitol from biomass is shown in Figure 4. It is a proven route already used in the preparation of xylitol syrups from residual cereal straws. The process begins with a physical treatment of biomass residues such as fine grinding or steam explosion to increase its surface area and accessibility to reagents (Chen et al. 2018).

The milled biomass is suitable for treatment with diluted sulfuric acid (5–6%) under boiling temperature for periods of about 1 hour. Hydrolysis promotes the cleavage of the xylan chain bonds, releasing xylose contaminated with a little glucose and mannose. It is a very important step because the yield at the end of the process depends directly on the concentration of the initial xylose solution. By neutralizing the acid with sodium hydroxide to neutrality, sodium sulfate pentahydrate can be recovered by simply cooling the solution to –20°C, as a precipitate. The sugars obtained are a mixture of C5 and C6 where xylose is predominant (see the HPLC chromatogram in Figure 1). Purification of xylose requires removal of the contaminating C6 sugars from the solution, glucose and mannose, to obtain xylose in the purest form possible. In the literature for this purification, methods such as ion exchange chromatography are considered, the cost of the process

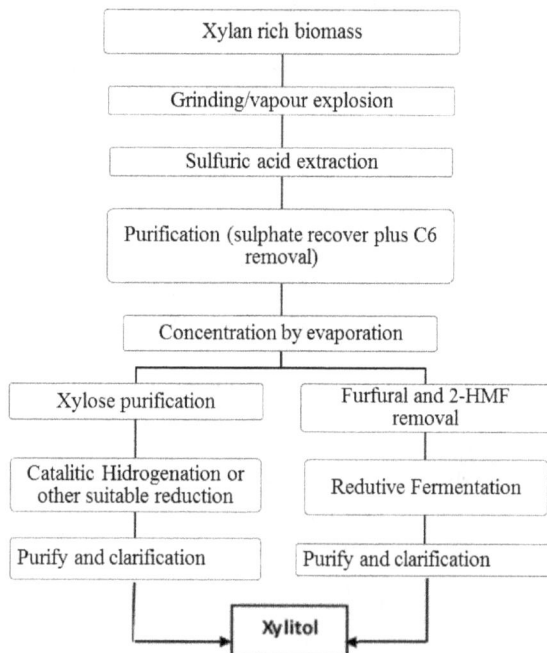

Figure 4. Diagram showing the possible ways to obtain xylitol from biomass.

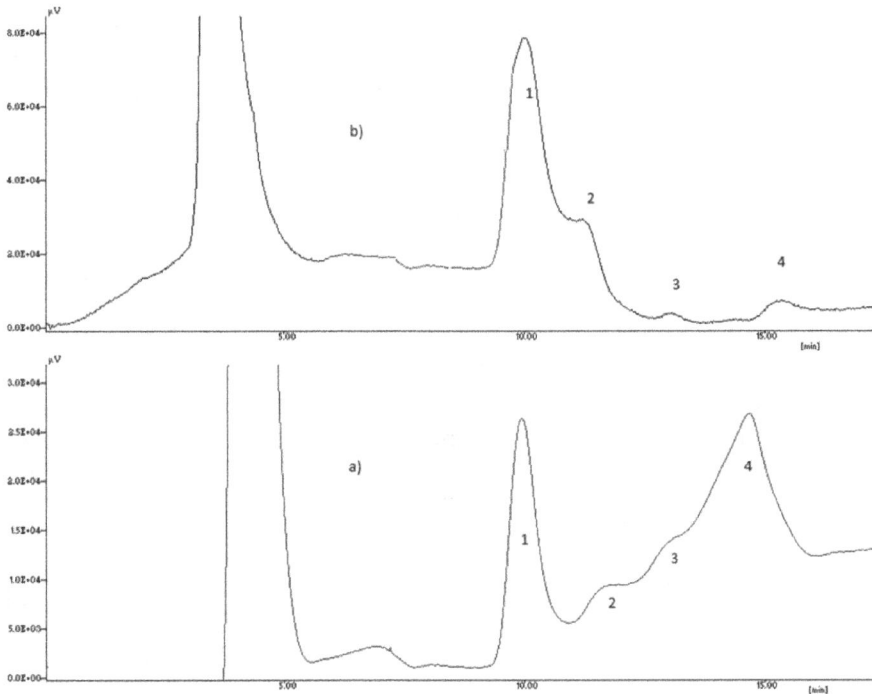

Figure 5. HPLC chromatograms showing a sugar composition of the neutralized corn cob hydrolyzate; (a) before inoculation and growth of *S. Cerevisiae* and (b) after 48 h of fermentation at 25°C. The bands corresponding to the sugars in C6 (3 and 4) dropped to residual levels, resulting in the end a purified xylose solution (band 1).

being important in the project. A much simpler and lower cost method can be used and consists of a simple fermentation of the xylose extracts with *S. Cerevisiae*, after simple pH correction to values above 3. As the yeast only metabolizes sugars in C6, it leaves C5 sugars intact by performing a clean removal of both hexoses. At the same time, partial removal of mineral salts occurs, and the xylose solution is purified by simple filtration. Figure 5 shows the HPLC chromatograms showing the removal of C6 sugars from biomass hydrolysates using the yeast *S. Cerevisiae*.

The third step consists of reducing xylose to xylitol, where different conversion paths can be followed: chemical, electrochemical or biotechnological, shown in Figure 4.

6.4.3 *Reduction of xylose by chemical/electrochemical methods*

A variety of chemical reducers can be used to convert xylose to xylitol. Both sodium borohydride (NaBH4) and aluminum lithium hydride (LiAlH4) are effective, but lead to the addition of undesirable ions to the product, thus requiring further purification. Either method generates xylitol solutions which, after ion exchange chromatography or dialysis, will give purified xylitol solutions.

A particularly interesting and easy carbonyl reduction can be implemented by electrochemical methods with anodic reduction with graphite electrodes, under conditions of controlled electrical potential to prevent water hydrolysis. Anaerobic conditions are required due to the easy oxidation of xylose to xylonic acid originating interference substances in the final product. Very pure xylitol can be produced with minimal energy expenditure and simple hardware using this method.

The hydrogenation process is a practical and easy alternative to reduce xylose to xylitol, since it does not generate any by-products or increases the ionic strength of the medium with the addition of any undesirable ionic species, as occurs with chemical reducers. The process uses hydrogen at low to medium pressures (PH2 = 3–50 atm), in the presence of a catalyst. The most used catalyst is Raney nickel, although several other hydrogenation catalysts can be used, such as noble metal salts (Pt, Pd, Ru). The last stage of production consists of isolation/purification and eventually crystallization of xylitol. The yield of the chemical process, as well as the quality of xylitol, are dependent on the purity of the initial xylose stream, since the presence of impurities interferes with the reduction process. In global terms, hydrogenation is a relatively simple and desirable technique, often a method to be chosen in the production of xylitol.

6.4.4 *Reduction of xylose by biotechnological techniques*

Biotechnological reductions are possible with the discovery of yeasts and other microorganisms capable of metabolizing D-xylose by reduction (Wang et al. 2013). This can be carried out by yeasts, bacteria, fungi or using enzymes purified from these microorganisms. The biotechnological route presents an alternative to the chemical route for obtaining xylitol. One of the advantages of using microorganisms is the concomitant removal of normal contaminants mannose/glucose sugars, which appear in hydrolysates due to their presence of hemicellulose or by partial hydrolysis of amorphous parts of cellulose.

Microorganisms that are important to identify are those capable of reducing xylose without being able to metallobolize it. The yield in some cases is low, and studies have been dedicated to the development of genetically engineered strains capable of reducing xylose more efficiently. In the selection of strains, since not all microorganisms can convert xylose, variables such as tolerances to high xylose concentration and the rate of processing reaction are important. The abiotic factors that influence the process and the operational conditions, the most relevant being the pH and temperature values and the availability of oxygen, but also the composition of the culture medium, such as the concentration of xylose and other monosaccharides, the presence of inhibitors, are evaluated (Wang et al. 2013).

Among the microorganisms already studied, yeasts that have excelled in the production of xylitol, are mainly *Candida guilliermondii*, *Candida subtropicalis*, *Candida tropicalis* and *Debaryomyces hansenii* (Li and Zhang 2016). The metabolism of xylose in yeast begins with its transport across the cell membrane by different mechanisms, since inside the cell, xylose is reduced to xylitol by the enzyme xylose reductase with the participation of the coenzyme NADPH or NADH (Wang et al. 2013).

However, the process is not very competitive in relation to catalytic hydrogenation, it takes longer due to the fermentation time, and the process tends to be more problematic and affected by undefined factors that are difficult to control by biotic factors that influence yeast growth and limiting fermentation results. The acid hydrolysis stage is also problematic because it gives rise to contaminants such as furfural and 5-hydroxymethylfurfural (2-HMF), the result of the degradation of sugars, acetic acid from the hydrolysis of hemicellulose acetyl groups and phenols from degradation of lignin. In biotechnological reduction, the presence of the above aldehydes has a major adverse effect on the specific growth rate of the microorganism. Their presence may require preliminary purification/detoxification techniques (Naidu et al. 2018), which reduces productivity. Many studies have been carried out in an attempt to find microorganisms resistant to these inhibitors, such as variants of *Saccharomyces cerevisiae*, and xylose fermenting yeasts, such as *Candida shehatae* and *Pichia stipitis*, which may be more resistant to the presence of the substances (Saleh et al. 2014).

6.4.5 *Finishing techniques in the production of xylitol*

Depending on the final form in which xylitol is to be produced, it must be subjected to purification and eventually crystallization steps. The removal of inorganic impurities consists of passing through ion exchange resin columns, clarification if needed with activated carbon treatment, simultaneously being capable of removing inorganic salts and adsorbing small phenolic pigments originating from lignin, generally giving to the extracts a slightly yellow to brown color. Ultrafiltration can be used to remove traces of protein and improve the performance of ion exchange resins. Often, the combination of several of these methods is used to improve the purity and color of xylitol.

Xylitol that is intended for sale on the market must undergo a crystallization process where part of it may be lost. Several methods can crystallize xylitol from the concentrated solution, almost all use crystallizers working at low temperatures with losses of 50% to 60%. However, for end use as a chewing gum or for the toothpaste industry, xylitol can simply be sent in the form of evaporated concentrate, in the form of a syrup, which ever is convenient for both the producer and the user.

7. Conversion of hemicellulose to furfural

The presence of furfuraldehyde was previously defined as undesirable as it constitutes an obstacle to fermentation with the use of yeasts when the ultimate goal is the production of ethanol or the reduction of xylose to xylitol. However, if a significant part of the biomass can be converted into furfural, this substance has use in different fields and thus has commercial value. No other source is known for furfural preparation other than the use of biomass. In fact, in 2019 the production of furfuraldehyde reached values close to 800,000 tons, with the main producers being China and South Africa. Other countries such as Colombia and India have lower production levels.

Much research has been done in recent years in an attempt to improve the process (Chompoopitch et al. 2017, Delbecq et al. 2018). The most common process used to obtain HMF depends on the treatment with sulfuric acid under conditions of high severity, when compared to those used when the objective is to recover the monosaccharides. The treatment of biomass with water vapor at temperatures up to 170°C, in a pressurized vessel with recirculation of the vapors in condenser allows the recovery of furfural. The yield can reach values close to 50%, therefore fraction other than hemicellulose starts to be processed, the most probable is the partial conversion of glucose, which generates some contamination with 2-hydroxymethyl furfural and other phenolic compounds derived from the hydrolysis of lignin (Zhicheng et al. 2019).

Xylose and hemicellulose in a high degree of purity were used as a study model for the preparation of furfural. The data obtained when using hemicellulose or xylose as starting material, gave results close to the total conversion, reaching values as high as 73–98% (Cuervo et al. 2020). From biomass, lower yield values are possible. As seen before, some xylans or other types of hemicellulose have connection points with both cellulose and lignin, positions that hinder hydrolysis of sugars. Hemicellulose rich in C6 sugars, such as mannose and glucose, is more difficult to hydrolyze and convert to 2-hydroxymethyl furfural. The attempt to improve the methods to do so includes their preparation in two stages: a first removal of hemicellulose leaving cellulose as a by-product for which end uses can easily be found, and then the hemicellulose fraction is subjected to conversion to HMF. Other aspects under study, such as the nature of the acid catalyst, both in the homogeneous and in the heterogeneous phase, like Bronsted or Lewis are in progress. The process can also use solvents, or mixtures of them, single-phase, two-phase, ionic liquids and a mixture of these different types.

Different types of biomass containing a high content of xylan-type hemicellulose can be conveniently transformed into furfural. Examples of substrates that are within this group are the ears of corn, the stalk of corn, and wood from soft trees, among others. As previously mentioned, xylan is poorly stable in an acidic environment, initially giving rise to water elimination reactions that eventually break the β(1,4) glycosidic bond, as depicted in the mechanism shown in Figure 6. Once carbon 1 of xylose is involved with carbon 4 of the next unit in the formation of a glycosidic bond, thus lowering its reactivity, acid protonation of hydroxyl groups at positions 3 or 4 are easily involved in water elimination, forming a vinyl alcohol which quickly rearranges into a mixture of ketones at positions 2 or 3 of the pyranose ring (sequence 1–3, see Figure 6). The carbonyl group formed activates the reactivity of its α carbons, promoting condensation reactions

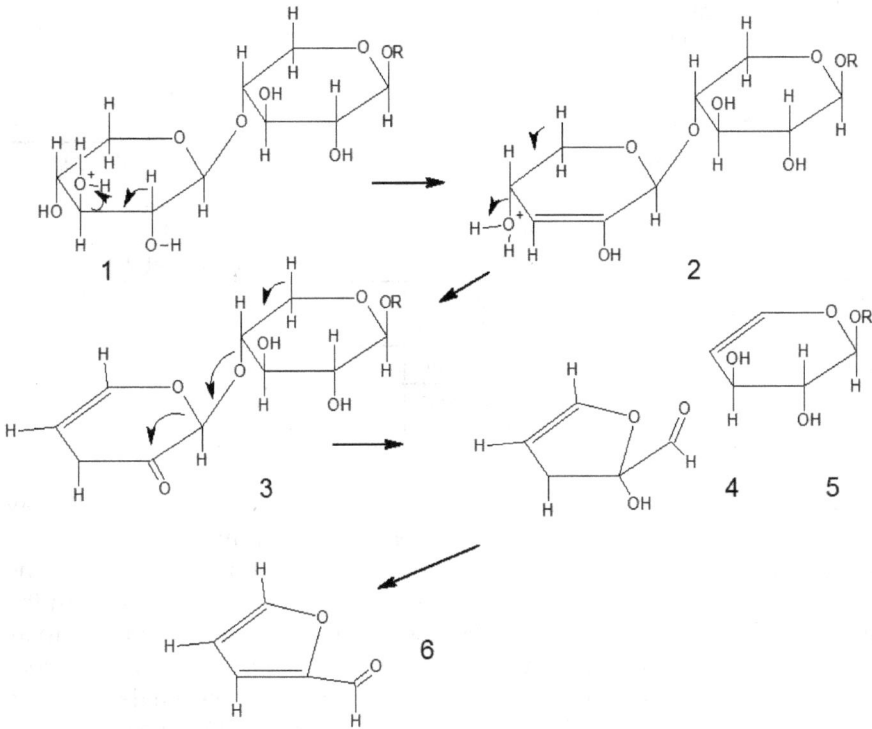

Figure 6. Possible acid-catalyzed hydrolytic decomposition mechanism of xylan in the generation of furfural.

and/or aldolic rearrangements. Following 3–4 steps in Figure 6 is an example of breaking bonds facilitated by the involvement of the glycosidic bond with the formation of a carbonyl group, which leads to the formation of furfural.

Commercial furfural is prepared mainly from biomass residues from existing processes, such as sugarcane bagasse, using acids and temperature under steam distillation, to quickly remove furfural from the acidic medium. The yields are generally less than 40%, as to be expected since hemicellulose fraction in biomass has a maximum value in that range, and constitutes the source of precursor sugars. The residue left from the furfural process can often be recycled for energy production. The problems associated with this process are acid effluents rich in lignin which, if not treated properly, can cause environmental problems (Dalvand et al. 2018). In Table 3, gives a compilation of HMF preparation results, with the respective experimental dehydration conditions.

Whatever the conditions used to promote triple dehydration of pentoses in HMF, its low stability means that it can undergo different condensation and addition reactions. Therefore, it is recommended to remove it quickly from the reaction medium and avoid contact with oxygen. The distillation with water vapor is suitable, as previously mentioned, or otherwise, the use of a

Table 3. Compilation of xylose and conversion of different biomass sources to hydroxymethylfurfural (HMF), using various experimental conditions.

Substrate	Conditions			Yield	Reference
	Solvent	Catalyst	T (°C)	(%)	
Xylose	γ-Valerolacton GLV/H2O (10%)	H2SO4	170	80	Gallo et al. 2013
Xylan	GLV/H2O	Al2(SO4)3	130	88	Yang et al. 2017
Corncob	H2O	HUSY zeolite	170	99	Li et al. 2017a
Sugarcane bagasse	H2O	Diluted H2SO4/NaCl	200	30	Hayelom et al. 2015
Corn stover	GVL	SC-CaCt-700	200	93	Li et al. 2017b
Rice straw	DMSO/H2O	HSO3-ZSM-5 zeolite	140	54	Hoang et al. 2020

process such as vacuum evaporation with subsequent condensation in low temperature trap can be considered, among others. From furfuraldehyde, a wide range of substances can be obtained by chemical transformation. The furfural reduction product, hydroxymethylfuran, is used as a solvent in the industry. The possibility of using HMF as a fuel is often suggested, due to its good burning properties in combustion engines that are already known from other substances related to furan. A recent review includes the detailed use of derivatives available in the literature (Mariscal et al. 2016). The production of furfural can be very interesting in a gradual biomass conversion process, especially as the final product is a material rich in cellulose and lignin, with its partially destroyed structure, capable of being separated and processed in order to obtain other bioproducts or biofuels.

8. References

Anonymous. 2010. Danisco launches XIVIA (TM). Food Australia 62: 555–555.
Ban, J., Shabbir, S. and M. Lim. 2017. Synthesis of L-ribose from D-ribose by a stereoconversion through sequential lactonization as the key transformation. Synthesis-Stuttgart 49: 4299–4302.
Bilal, M., Iqbal, H. M. N. and H. Hu. 2019. Metabolic engineering pathways for rare sugars biosynthesis, physiological functionalities, and applications—a review. Critical Reviews in Food Science And Nutrition 58: 2768–2778.
Chadni, M., Bals, O. and I. Ziegler-Devin. 2019. Microwave-assisted extraction of high-molecular-weight hemicelluloses from spruce wood. Comptes Rendus Chimie 22 574–584.
Chatani, N., Tokuhisa, H., Kokubu, Satoru F. and M. Shinji. 1995. CO-2(CO)(8)-catalyzed reaction of aromatic-aldehydes with hydrosilanes under carbon-monoxide as 1 Atm—incorporation of CO into the carbonyl carbon atom of aldehydes. Journal of Organometallic Chemistry 499: 193–197.
Chen, X., Li, H. and S. Sun. 2018. Co-production of oligosaccharides and fermentable sugar from wheat straw by hydrothermal pretreatment combined with alkaline ethanol extraction. Industrial Crops and Products 111: 70–77.
Chompoopitch, T., Vorranutch, I., Siripit, S., Busaya, C., Supawadee, N., Kajornsak, F., Tawatchai, C., Radchadaporn, K., Nanthiya, H., Noriaki, S. and H. Hirofumi. 2017. Dehydration of

D-xylose to furfural using acid-functionalized MWCNTs catalysts. Adv. Nat. Sci: Nanosci. Nanotechnol. 8: 3.

Chowdhury, M. D., Tazul, I., Naito, M. and R. C. Yanagita. 2015. Synthesis of 6-O-decanoyl-D-altrose and 6-O-decanoyl-D-gulose and evaluation of their biological activity on plant grow. Plant Growth Regulation 75: 707–713.

Cuervo, P., Romanelli, O. H., Gustavo, P., Cubillos, J. A., Rojas, H. A. and J. J. Martínez. 2020. Selective catalytic dehydration of xylose to furfural and fructose and glucose to 5-hydroximethylfurfural (HMF) using preyssler heteropolyacid. Chemistry Select 5: 4186–4193.

da Silva, E. G., Borges, A. S. and N. R. Maione. 2020. Fermentation of hemicellulose liquor from Brewer's spent grain using Scheffersomyces stipitis and Pachysolen tannophilus for production of 2G ethanol and xylitol. Biofuels Bioproducts & Biorefining-BIOFPR 14: 127–137.

Dall Cortivo, P. R., Hickert, L. R. and C. A. Rosa. 2020. Conversion of fermentable sugars from hydrolysates of soybean and oat hulls into ethanol and xylitol by Spathaspora hagerdaliae UFMG-CM-Y303. Industrial Crops and Products 146: 112218.

Dalvand, K., Rubin, J. and S. Gunukula. 2018. Economics of biofuels: Market potential of furfural and its derivative. Biomass & Bioenergy 115: 56–63.

Danisco Cuts Back Xylitol Business. 2009. Chemical & Engineering News 87: 14–14.

Dax, D., Chavez, B., Soledad, M. and C. Honorato. 2015. Tailor-made hemicellulose-based hydrogels reinforced with nanofibrillated cellulose. Nordic Pulp & Paper Research Journal 30: 373.

Delbecq, F., Wang, Y., Muralidhara, A., Karim, E. O., Guy, M. and L. Christophe. 2018. Hydrolysis of hemicellulose and derivatives—a review of recent advances in the production of furfural. Frontiers in Chemistry 6: 146.

Du, X., Perez-Boad, M., Fernandez, C., Rencoret, J., del Rio, J. C., Jimenez-Barbero, J., Li, J. B., Gutierrez, A. and A. T. Martinez. 2014. Analysis of lignin-carbohydrate and lignin-lignin linkages after hydrolase treatment of xylan-lignin, glucomannan-lignin and glucan-lignin complexes from spruce wood. 239: 1079–1090.

Espinoza-Acosta, J. L. 2020. Biotechnological production of xylitol from agricultural waste. Biotecnia 22: 126–134.

Esteves, B., Graca, J. and H. Pereira. 2008. Extractive composition and summative chemical analysis of thermally treated eucalypt wood. Holzforschung 62: 344–351.

Fonseca Silva, T. C., Habibi, Y., Colodette, J. L. and L. A. Lucia. 2011. The influence of the chemical and structural features of xylan on the physical properties of its derived hydrogels. Soft Matter 7: 1090.

Gallina, G., Cabeza, A. and P. Biasi. 2016. Optimal conditions for hemicelluloses extraction from Eucalyptus globulus wood: hydrothermal treatment in a semi-continuous reactor. Fuel Processing Technology 148: 350–360.

Gallo, J. M. R., Alonso, D. M. and M. A. Mellmer. 2013. Production of furfural from lignocellulosic biomass using beta zeolite and biomass-derived solvent. Topics in Catalysis 56: 1775–1781.

Gargouri, W., Zmantar, T. and R. Kammoun. 2018. Coupling xylitol with remineralizing agents improves tooth protection against demineralization but reduces antibiofilm effect. Microbial Pathogenesis 123: 177–182.

Gargouri, W., Kammoun, R., Elleuche, M., Tlili, M., Kechaou, N. and S. Ghoul-Mazgar. 2020. Effect of xylitol chewing gum enriched with propolis on dentin remineralization *in vitro*. Arch. Oral. Biol. 112: 104684.

Geng, W., Narron, R., Jiang, X., Pawlak, J. J., Chang, H. M., Park, S., Jameel, H. and R. A. Venditti. 2019. The influence of lignin content and structure on hemicellulose alkaline extraction for non-wood and hardwood lignocellulosic biomass. Cellulose 26: 3219–3230.

Giummarella, N. and M. Lawoko. 2017. Structural insights on recalcitrance during hydrothermal hemicellulose extraction from wood. ACS Sustainable Chemistry & Engineering 5: 5156–5165.

Grembecka, M. 2015. Sugar alcohols-their role in the modern world of sweeteners: A review. European Food Research and Technology 241: 1–14.

Hayelom, G., Kiros, F., Tsegalaul, K. and G. Tsigehiwot. 2015. Synthesis of furfural from bagasse. International Letters of Chemistry, Physics and Astronomy 57: 72–84.

Hoang, P. H., Dat, N. M., Cuong, T. D. and D. T. Tung. 2020. Production of 5-hydroxymethylfurfural (HMF) from rice-straw biomass using a HSO3-ZSM-5 zeolite catalyst under assistance of sonication. RSC Adv. 10: 13489–13495.

Hu, G. and W. Yu. 2013. Binding of cholesterol and bile acid to hemicelluloses from rice bran. International Journal of Food Sciences and Nutrition 64: 461–466.

Inokuma, K., Iwamoto, R., Bamba, T., Hasunuma, T. and A. Kondo. 2017. Improvement of xylose fermentation ability under heat and acid co-stress in Saccharomyces cerevisiae using genome shuffling technique. Frontiers in Bioengineering and Biotechnology 5: 81.

Jeong, S. M., Yong-Jin, K. and L. Dong-Hoon. 2012. Ethanol production by co-fermentation of hexose and pentose from food wastes using Saccharomyces coreanus and Pichia stipitis. Korean Journal of Chemical Engineering 29: 1038–1043.

Kamzolova, S. V., Shamin, R. V., Stepanova, N. N., Morgunov, G. I., Lunina, J. N., Allayarov, R. K., Samoilenko, V. A. and I. G. Morgunov. 2018. Fermentation conditions and media optimization for isocitric acid production from ethanol by Yarrowia lipolytica. Biomed Research International 2018: 2543210.

Koropatkin, N. M., Cameron, E. A. and E. C. Martens. 2015. How glycan metabolism shapes the human gut microbiota. Nature Reviews Microbiology 10: 323–335.

Kuhad, R. C., Gupta, R., Khasa, Y. P., Singh, A. and Y. H. P. Zhang. 2011. Bioethanol production from pentose sugars: Current status and future prospects. Renewable & Sustainable Energy Reviews 15: 4950–4962.

Le, B., Ngoc, A. P. T. and S. H. Yang. 2020. Synbiotic fermented soymilk with Weissella cibaria FB069 and xylooligosaccharides prevents proliferation in human colon cancer cells. Journal of Applied Microbiology 128: 1486–1496.

Li, Xiaoyun, Liu, Qingling and Luo, Chunhui. 2017a. Kinetics of furfural production from corn cob in gamma-valerolactone using dilute sulfuric acid as catalyst. ACS Sustainable Chemistry & Engineering 5: 8587–8593.

Li, W., Zhu, Y., Lu, Y., Liu, Q. Y., Guan, S. N., Chang, H. M., Jameel, H. and L. L. Ma. 2017b. Enhanced furfural production from raw corn stover employing a novel heterogeneous acid catalyst. Bioresource Technology 245: 258–265.

Li, H. and L. Zhang. 2016. Xylitol production from waste xylose mother liquor containing miscellaneous sugars and inhibitors: one-pot biotransformation by Candida tropicalis and recombinant Bacillus subtilis. Microbial Cell Factories 15: 82.

Liu, J., Chinga-Carrasco, G., Cheng, F., Xu, W. Y., Willfor, S., Syverud, K. and C. L. Xu. 2016a. Hemicellulose-reinforced nanocellulose hydrogels for wound healing application. Cellulose 23: 3129–3143.

Liu, X., Lin, Q. and Y. Yan. 2019b. Hemicellulose from plant biomass in medical and pharmaceutical application: A critical review. Current Medicinal Chemistry 26: 2430–2455.

Lynch, D. J., Michalek, S. M., Zhu, M., Drake, D., Qian, F. and J. A. Banas. 2013. Cariogenicity of streptococcus mutans glucan-binding protein deletion mutants. Oral Health Dent Manag. 12: 191–199.

Mäkinen, K. K. 2000. The rocky road of xylitol to its clinical application. Journal of Dental Research 79: 1352–1355.

Mariscal, R., Maireles-Torres, P., Ojeda, M., Sádaba, I. and M. López Granados. 2016. Furfural: A renewable and versatile platform molecule for the synthesis of chemicals and fuels. Energy Environ. Sci. 9: 1144–1189.

Mattila, P. T., Svanberg, M. J. and M. L. E. Knuuttila. 1998. Dietary xylitol protects against osseal changes in experimental osteoporosis. pp. 157–162. In: Burckhardt, P., Dawson-Hughes, B. and Heaney, R. P. (eds.). Nutritional Aspects of Osteoporosis. Proceedings in the Serono Symposia USA Series. Springer, New York, NY.

Maurya, D. P., Singla, A. and S. Negi. 2015. An overview of key pretreatment processes for biological conversion of lignocellulosic biomass to bioethanol. Biotech. 5: 597–609.

Naidu, D. S., Hlangothi, S. P. and M. J. John. 2018. Bio-based products from xylan: A review. Carbohydrate Polymers 179: 28–41.

Nishimura, S., Yajima, K., Harada, N., Harada, N., Ogawa, Y. and N. Hayashi. 1994. Automated synthesis of radio pharmaceuticals for pet—an apparatus for [1-C-11] labelled aldoses. Journal of Automatic Chemistry 16: 195–204.

Park, E., Na, H. S., Kim, S. M., Wallet, S., Cha, S. and J. Chung. 2014. Xylitol, an anticaries agent, exhibits potent inhibition of inflammatory responses in human THP-1-derived macrophages infected with Porphyromonas gingivalis. Journal of Periodontology 85: 212–223.

Pikis, A., Hess, S., Arnold, I., Erni, B. and J. Thompson. 2006. Genetic requirements for growth of Escherichia coli K12 on methyl-alpha-D-glucopyranoside and the five alpha-D-glucosyl-D-fructose isomers of sucrose. Journal of Biological Chemistry 281: 17900–17908.

Rao, R. G., Ravichandran, A. and G. Kandalam. 2019. Screening of wild basidiomycetes and evaluation of the biodegradation potential of dyes and lignin by manganese peroxidases. Bioresources 14: 6558–6576.

Rehman, S., Nadeem, M. and F. Ahmad. 2013. Biotechnological production of xylitol from banana peel and its impact on physicochemical properties of rusks. Journal of Agricultural Science and Technology 15: 747–757.

Rodrussamee, N., Sattayawat, P. and M. Yamada. 2018. Highly efficient conversion of xylose to ethanol without glucose repression by newly isolated thermotolerant Spathaspora passalidarum CMUWF1-2. BMC Microbiology 18: 73.

Saleh, M., Cuevas, Manuel, Garcia, Juan F. and S. Sanchez. 2014. Valorization of olive stones for xylitol and ethanol production from dilute acid pretreatment via enzymatic hydrolysis and fermentation by Pachysolen tannophilus. Biochemical Engineering Journal 90: 286–293.

Salli, K., Lehtinen, M. J. and K. Tiihonen. 2019. Xylitol's health benefits beyond dental health: A comprehensive review. Nutrients 11: 1813.

Sannigrahi, P., Ragauskas, A. J. and G. A. Tuskan. 2010. Poplar as a feedstock for biofuels: A review of compositional characteristics. Biofuels Bioproducts & Biorefining-BioFPR 4: 209–226.

Sauraj, Kumar, Uday, S. and P. Gopinath. 2017. Synthesis and bio-evaluation of xylan-5-fluorouracil-1-acetic acid conjugates as prodrugs for colon cancer treatment. Carbohydrate Polymers 157: 1442–1450.

Seo, E. S., Kim, D. and J. F. Robyt. 2004. Modified oligosaccharides as potential dental plaque control materials. Biotechnology Progress 20: 1550–1554.

Stahlberg, T., Woodley, J. M. and A. Riisager. 2012. Enzymatic isomerization of glucose and xylose in ionic liquids. Catalysis Science & Technology 2: 291–295.

Stepowska, H. and A. Zamojski. 1999. Elongation of the pentose chain at the terminal carbon atom with Grignard C-1 reagents. A study of the homologation reaction. Tetrahedron 55: 5519–5538.

Svard, A., Brannvall, E. and U. Edlund. 2015. Rapeseed straw as a renewable source of hemicelluloses: Extraction, characterization and film formation. Carbohydrate Polymers 133: 179–186.

Tiwari, Utsav P., Singh, Amit K. and J. Jha Rajesh. 2019. Fermentation characteristics of resistant starch, arabinoxylan, and beta-glucan and their effects on the gut microbial ecology of pigs: A review. Animal Nutrition 5: 217–226.

Vilela, L. F., Gomes de Araujo, V. P. and R. S. Paredes. 2015. Enhanced xylose fermentation and ethanol production by engineered Saccharomyces cerevisiae strain. AMB Express 5: 16.

Wang, L., Wu, D. and P. Tang. 2013. Effect of organic acids found in cottonseed hull hydrolysate on the xylitol fermentation by Candida tropicalis. Bioprocess and Biosystems Engineering 36: 1053–1061.

Wu, S., Dai, X., Kan, J. R., Shilong, F. D. and M. Y. Zhu. 2017. Fabrication of carboxymethyl chitosan-hemicellulose resin for adsorptive removal of heavy metals from wastewater. Chinese Chemical Letters 28: 625–632.

Yang, T., Zhou, Y. H., Zhu, S. Z., Pan, H. and Y. B. Huang. 2017. Insight into aluminum sulfate-catalyzed xylan conversion into furfural in a γ-Valerolactone/water biphasic solvent under microwave conditions. Chemsuschem. 10: 4066–4079.

You, S., Xie, C. and R. Ma. 2019. Improvement in catalytic activity and thermostability of a GH10 xylanase and its synergistic degradation of biomass with cellulase. Biotechnology for Biofuels 12: 278.

Yuan, Y., Zou, P. and J. Zhou. 2019. Microwave-assisted hydrothermal extraction of non-structural carbohydrates and hemicelluloses from tobacco biomass. Carbohydrate Polymers 223: 115043.

Zhicheng, J., Remon, J., Li, T., Fan, J. J., Hu, C. W. and J. H. Clark. 2019. A one-pot microwave-assisted NaCl-H2O/GVL solvent system for cellulose conversion to 5-hydroxymethylfurfural and saccharides with *in situ* separation of the products. Cellulose 26: 8383–8400.

Chapter 3

Biomass Delignification with Biomimetic Enzyme Systems

1. Introduction

Lignin is one of the most abundant natural polymers after cellulose and hemicellulose. It gives the wood its natural brown color and with its removal, wood without lignin, has different properties and has an important discoloration, allowing for the preparation of a range of new derived products—transparent wood is a recent example (Wu et al. 2019). The removal of lignin from biomass, while retaining the remaining components, can give rise to materials with high mechanical strength and density, capable of preparing new composites. These materials can be subjected to different types of carbonization to produce new allomorphic carbon materials. Treatments that have successfully removed lignin from biomass use mimetic biocatalysts constructed with metalloporphyrins (Zucca et al. 2014, Xie et al. 2018) coupled with oxidation steps of the Baeyer-Villiger type, a process that can be optimized by the use of especially robust porphyrins (Fang et al. 2018). The delignification products are vanillic acid or its esters, which can be recovered, and there are uses for them (Zhang et al. 2020).

One of the intrinsic characteristics of lignin due to its polymeric nature of phenylpropane units is its high stability to degradation. Some natural fungi have developed specific enzymes—ligninases that are capable of degrading lignin. These fungi are generally capable of processing aged biomass, in which the most labile components—hemicellulose and amorphous cellulose are already partially hydrolyzed and absent.

This allows for easy entry of enzymes to biomass sites otherwise blocked by the highly branched hemicellulose structure.

The best knowledge of ligno-cellulosic enzymes comes from their isolation and characterization of the extracellular medium of cultures of *Phanaerocaete crysosporium* fungi. Two enzymes essential to the process are lignin peroxidase (LiP) and manganese peroxidase (MnP), both containing a protoporphyrin IX iron complex as a prosthetic group. LiP has an active site with an electron donor N atom of a histidine or tryptophan amino acid residue coordinated with the central metal as the fifth ligand (proximal position). The cytochrome P-450 enzyme family is known to have a similar prosthetic group. In the distal position, another residue containing an N atom stabilizes the reaction intermediates. The necessary cofactor is hydrogen peroxide, however, in some cases, oxygen has proven to serve as a substitute as an oxygen donor for the formation of high valency, very high oxidizing oxo reaction intermediates. Manganese peroxidase (MnP) does not contain a distal histidine or its position is modified, it is thought to allow the approximation of Mn2+ ions to metalloporphyrin, and to be oxidized to the Mn3+ species (Hofrichter 2002), a potent one electron oxidizer used as a mediator in the transfer of free radicals to lignin, in the form of a complex with ligands available as lactate [Mn3+ (lactate)] or malonate (Hilden et al. 2014). This coordination may facilitate the passage of the free radical through the channels of carbohydrate polymers to susceptible lignin positions (benzylic positions).

Over the years, many biomimetic models of cytochromes and ligninases have been developed, made possible by the evolution of synthetic routes for the preparation of the elements of the active site enzyme namely the porphyrinic ligand (Xie et al. 2018). This requires the preparation of metalloporphyrin with the minimum characteristics necessary to imitate the functioning of the enzyme cycle, certainly with a lack of regulatory mechanisms and possible "defects". The main elements are metalloporphyrin, the fifth ligand in the axial position, such as histidine or a similar substance such as pyridine or imidazole, which is found to be a key element that must be provided to activate the reactivity of the catalyst, and a donor of oxygen, such as hydrogen peroxide (Lu et al. 2019).

Biomimetic systems developed with porphyrins such as those found in enzymes, such as porphyrins substituted with alkyl groups in the β-pyrrolic positions (see Figure 3) proved to be not stable to the reaction conditions, being quickly destroyed by the reaction intermediates generated in the catalytic cycle, possessing a very high oxidative potential. The evidence of these "naked" porphyrins undergoing degradation very easily is clearly seen in the reaction medium by the rapid disappearance of the strong brown color of the porphyrin metal complexes. Although natural ligninases have a similar metal complex, that is, flat porphyrin macrocycles such as protoporphyrin IX, they reside in a protective cage of a protein nature. The synthetic models are designed to try to simulate the same protection by adding bulky groups or atoms in the ligand, so the porphyrinic ligand must have these structural requirements.

Metalloporphyrin inactivation occurs through different degradation pathways; its understanding allows the design of suitable porphyrin ligands (see below). The synthesis of metalloporphyrins is a multi-step process, some with low yield, which makes its preparation challenging. Its high structural strength is necessary to guarantee a high turnover of catalytic reaction and stability so that the biomimetic process is viable and does not lead to the immediate disappearance of the catalyst. For this, structural characteristics in the porphyrins are necessary to prevent its destruction.

2. The catalytic cycle of delignification

In order to understand the challenges imposed to the metal complexes found in the peroxidase structure, an analytical discussion about the delignification mechanism must be done. The accepted mechanism of the catalytic cycle of ligninases is similar to that observed in enzymes of the cytochrome P-450 family, in which the porphyrin iron (II) complexes, in the presence of an oxygen donor, form a high valence oxoferril complex with radical nature (see Figure 1).

The first stage involves the coordination of the central metal Fe (II) of porphyrin (at top left in Figure 1) with hydrogen peroxide to produce an iron peroxide (III) complex (Cpd 0), which by rearrangement produces the intermediate oxo iron (IV) radical cation (Cpd 1). This species assumes

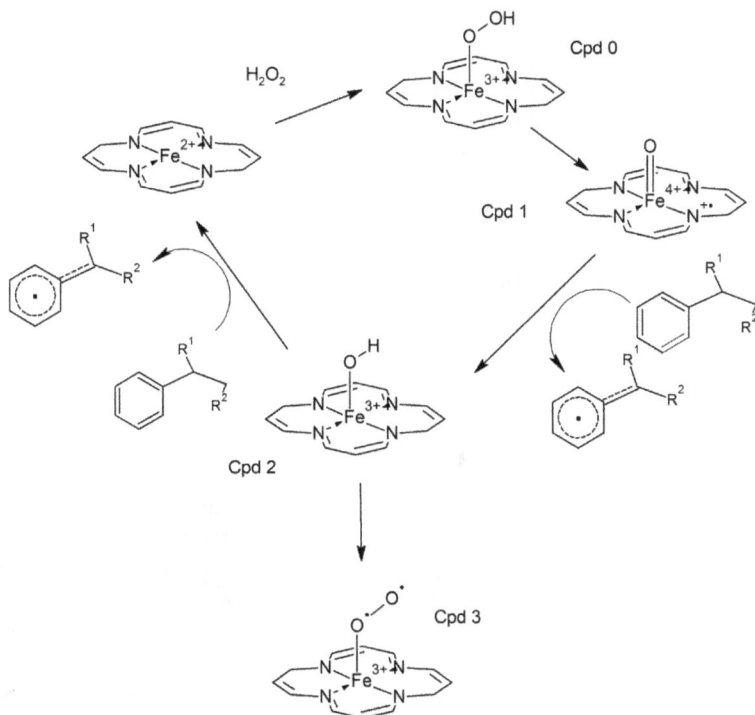

Figure 1. Mechanistic pathway accepted for ligninases. An iron (II) complex reacts with hydrogen peroxide to form a ferric peroxo derivative (Cpd 0) that, by rearrangement, produces the oxoferril complex (Cpd 1). This species is a strong oxidizer capable of abstracting H radicals from the benzylic positions of lignin, resulting in the formation of Cpd 2. This has sufficient oxidative capacity to repeat the process, ultimately regenerating the initial ferrous metalloporphyrin. Cpd 2 in the presence of excess hydrogen peroxide can give rise to the so-called Cpd 3, with little oxidative capacity.

different forms of resonance that involve the ligand and the ferryl oxygen atom. The resonance forms can be written as Porph(Fe V)=O, where the iron atom that presents the oxidation state +5 (high valence); Porph+•(Fe IV)=O in the case of a transfer of an electron from the aromatic ligand resonance bonds of porphyrin to the ferryl atom; and Porph(FeIV)–O• when one of the double oxo iron bond breaks homolytically giving rise to an electron resonating in the aromatic structure. The oxygen atom attached to the central iron, in this last form, has a high oxidizing reactivity, capable of abstraction of the radical hydrogen (H •) of any nearby species. The oxidizing capacity of that species is so high that it is considered one of the few species capable of oxidizing primary carbons from low reactive alkanes, such as methane, as can be seen in methane monooxygenases MMO's (Kudrik et al. 2012). The reverse of this high reactivity are the catalyst self-destruction problems observed (da Silva et al. 2015), mainly at the level of the ligand positions accessible to oxo oxygen either from Cpd 1 or Cpd 2. Porphyrins have resonance forms that lead to transient conformations outside the planarity, favoring self-oxidation. The presence of a fifth ligand coordinate basic ligand, such as imidazole or pyridine base, is necessary to contribute to stabilize this species, in association with the aromatic porphyrin ring with its 20π electrons. The same effect occurs in natural enzymes, the use of ligninases in industrial processes requires the presence of veratrilic alcohol (3,4-dimethoxybenzyl alcohol) as a protective substance to prevent the self-destruction of the central metalloporphyrin of the enzyme and allow greater enzymatic turnover, with the alcohol oxidized to the corresponding aldehyde (Mathew et al. 2015).

If the Cpd 1 complexes have access to lignin, usually surrounded by low reactivity sugars, the abstraction of an H • from an α (benzylic position) or β carbons from the lignin phenyl-propane structure is expected. This leads to the formation of benzylic radicals and the formation of Cpd 2 (see Figure 1). This last species has less oxidative capacity than Cpd 1, but enough energy to repeat the process of an abstraction of H • from another phenylpropanoid chain to produce the initial state of porphyrin (see Figure 1). Cpd 2 in the presence of high concentrations of H2O2 can form another intermediate, with little oxidative capacity, known as Cpd 3 (see Figure 1). This species has peroxide oxygen, which may give rise to reactions that lead to inactivation of the metalloporphyrin complex. This information is especially relevant in the design of construction of biomimetic systems, where the transient existence of high concentrations of peroxide can originate in the initial stages of the reaction, when adding it (see below).

The structure of the phenylpropanoid lignin resin is very diverse but is prepared from the radicalar polymerization of phenylpropane monomers. In the alkylic portion of phenylpropane, several types of chemical bonds between lignin monomers can be found. Essentially two types are found: alkyl-aryl ether bonds in the α or β positions of phenylpropane, or a carbon-carbon bond in the same alkyl part of the phenolic monomers. In both situations, the benzylic group is always present in the phenylpropanoid fraction (see Figure 2, 1). Nature prepares lignin from vinyl monomers through

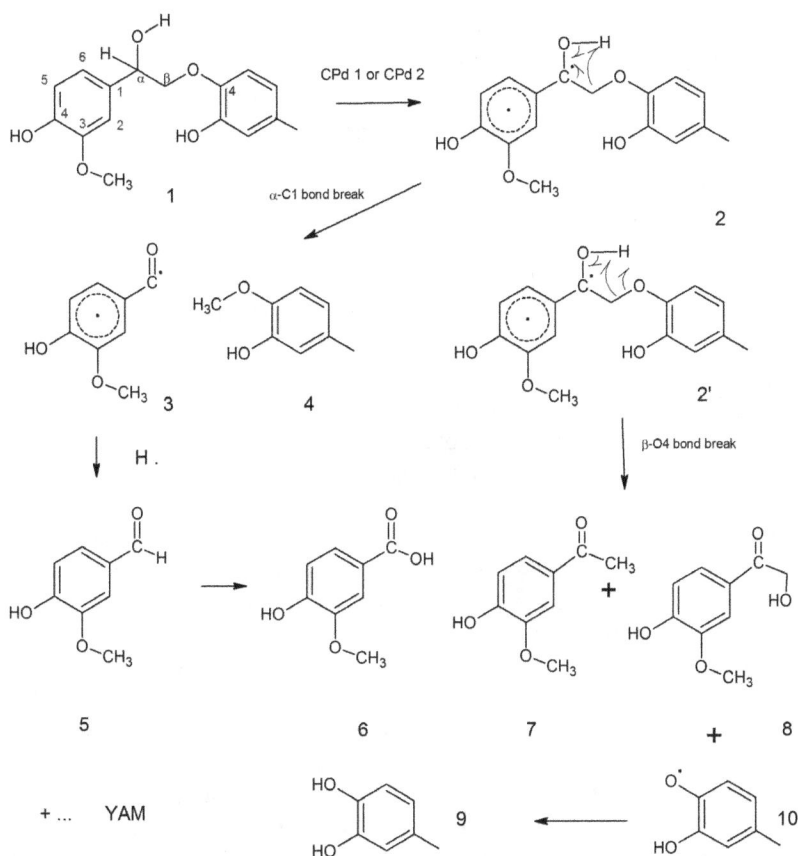

Figure 2. Examples of lignin chain breakage induced by the presence of Cpd 1 and Cpd 2, originated in the activity of ligninases. The break in the Cα-Cβ bond leads to the formation of benzaldehyde derivatives and is shown in steps 2–6. An alternative mechanism that leads to the cleavage of the C-O β-ether bond is shown in sequence 2' to 7, 8 and 9, 10. The result is the "yield of aromatic molecules" or aromatic molecule yield (YAM).

polymerization with free radicals, but interestingly originating a polymer rich in benzylic groups, known for their characteristic to strongly stabilize free radicals. From these, degradation products of the polymer can be inferred.

Thus, the presence of benzylic radicals constitutes the initial step of lignin depletion. Either Cpd 1 or Cpd 2 abstract H • radicals from phenolic or alkyl lignin positions generating a radical that quickly resonates by rearranging by abstraction of nearby hydrogen radicals. It rearranges the radical in the benzylic radical, due to its greater stability. This characteristic of the "radical trap" type gives lignin high antioxidant properties, known and described (Li et al. 2020b). The reactivity of the benzylic radical is well understood and is one of the positions where a bond break is likely to occur in the lignin chain. One of the possible pathways, through the break in the bond between the α and β carbon of the propane part (steps 1–4, see Figure 2). This break gives rise to a benzylic carbonyl radical generating as a final product a derivative of benzaldehyde (5) by abstraction of H •, one of the intermediates found in the delignification process (conversion 2 to 3, see Figure 2). Aldehydes are easily oxidized by atmospheric oxygen, giving rise to the corresponding carboxylic acids, such as (6), common found among the final products of the delignification process. However, radical species are very reactive and other non-specific products can be found. An alternative mechanism leads to the breakdown of the β-O-4 ether bond (conversion of 2' into 7, 8 and 10, see Figure 2), leading to the formation of compounds with the ketone group from the benzylic position, as examples 7–8. These processes result in the transformation of the lignin resin structure into a large number of small phenolic fragments that easily diffuse out of the hemicellulose/lignin domain. Thus, the final product of lignin decomposition is a set of aromatic substances of low molar mass, related to phenols, whose yield is usually called YAM (yield of aromatic molecules; Witzler et al. 2018). Experimental conditions that favor the formation of specific substances of the type of products, but in higher yield have been the object of study in recent years. Some of these substances, associated with the oxidative decomposition of lignin, are vanillic acid and its esters, such as methyl 5-carbomethoxyvanylate.

Oxidative decomposition mechanisms similar to these, by the action of free radicals, can be found in the degradation of other types of molecules, such as pharmaceutical substances. An example will be diclofenac (DCF), which through oxidative degradation induced by reactive oxygen species (ROS) formed from the activation of oxygen photo-excitation by irradiation by mediation with photosensitizers (Diaz-Angulo et al. 2019) or by water oxidation by holes (h+), when in contact with solid electrically conductor catalysts, able to form e−/h+ pairs, the last a species capable of generating reactive oxygen species (ROS) from water, such as HO•, leading to the decomposition of DCF into low molecular weight fragments (Silvestri et al. 2019). Biomass has no h+ conductivity capacity and, therefore, a catalyst capable of forming radical species in proximity to the lignin domain is the solution, as Nature does with ligninases.

3. Biomimetic systems

Biomimetic systems are simple imitations of enzymes in which only the essential characteristics of the active enzyme center are predicted, while maintaining the ability to perform the catalytic cycle. In the present case, biomimetic catalysts will have to carry out the joint actions represented in Figures 1 and 2. In these simpler models, the lack of apoprotein causes multiple mechanisms of enzymatic specificity and regulation to be absent, together with stereo protection of cavities made with protein around the metallocomplex of the active center. In any case, if they are properly constructed, they may show a high catalytic activity, without the problems of regulation and/or inhibition, in the present case in delignificant oxidation functions, mixed LiP and MnP functionality. Additionally, the lack of apoprotein tends to make these catalysts smaller, which may be an advantage in the treatment of biomass where hemicellulose is the place to be reached, since it has defenses against the entry of enzymes, but not in relation to small molecules. The cofactors available for the catalytic cycle are already present or can be transported together with the catalysts, when they are diffused. Thus, it is to be expected that these biomimetic systems are capable of delignifying biomass, under convenient low temperature reaction conditions.

3.1 Stability of porphyrin ligands to high valence oxo metal complexes

The main problem with the simplest biomimetic models of delignification is the stability of the catalyst to the exceptionally oxidation reactive intermediates formed in the catalytic cycle of ligninases, especially Cpd 2. The use of β-pyrrolic-substituted porphyrin metal complexes, such as protoporphyrin IX (1, see Figure 3) or substituted meso-alkyl (3 in Figure 3), results in the rapid destruction of the corresponding metalloporphyrins, that is, the aromatic macrocycle by different types of mechanisms, some that inactivate the catalyst, others that lead to the destruction of the ring. These porphyrins can, however, be used as long as they are immobilized on a support capable of providing a cage-like structure (Ji et al. 2019).

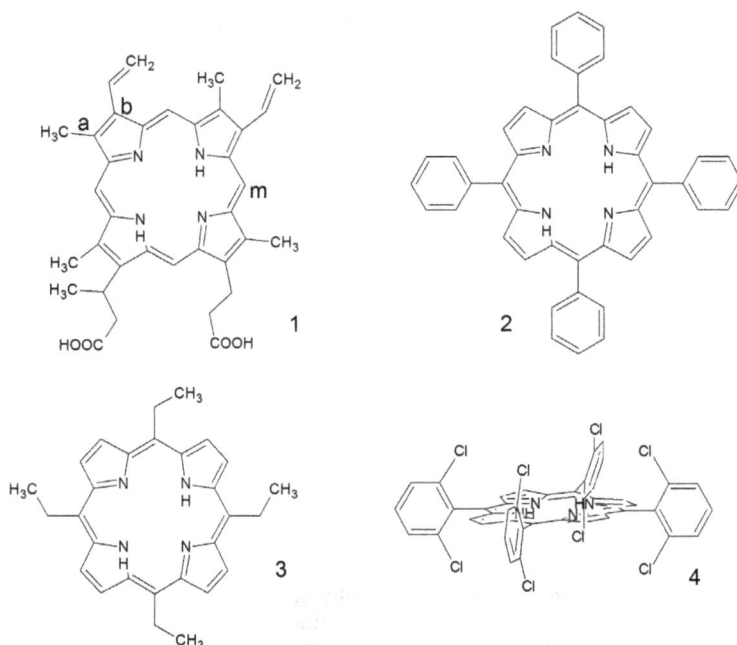

Figure 3. Structure of the porphyrin ligand used in natural systems and tested in artificial systems: 1-Protoporphyrin IX, 2-meso tetrakis phenyl porphyrin, 3-meso tetrakis ethyl porphyrin, 4-meso tetrakis (2,6-diclophenyl) porphyrin or TDCPP. Structure (4) is one of the most stable ligands known in the construction of biomimetic ligninase systems. The β-pyrrolic positions in the macrocycle are assigned by the letters a and b in (1), one of the four meso positions per m.

The main routes of inactivation of the active center of biomimetic catalysts are shown in Figure 4. The access of the catalysts to the substrate is essential to the good stability of the catalyst. If this is not possible, high-valence oxo species will have a longer life span, giving them the opportunity to carry out intramolecular reactions that result in their degradation or the formation of inactive complexes (Banfi et al. 1988). These pathways degradation can occur during the normal catalytic cycle and lead to the destruction of the catalyst.

One of the pathways, especially if the propane part is sterically blocked in the lignin, an already formed lignin radical intermediary may approach and covalently bond to one of the porphyrin nitrogen leading to N-alkylation of the macrocyclic ring, resulting in a complex with the active center irreversibly covalently linked to lignin and thus blocked (path i, Figure 4). This phenomenon is characteristic of the chemical activity of Cytochrome P-450, for example in the oxidation of sterically blocked olefins in a process generally referred to as N-suicidal destruction (Manno et al. 1995).

Another important degradation pathway is the formation of the μ-oxo dimer (path ii in Figure 4) in which an oxo-ferryl complex (Cpd 1 or Cpd 2) oxidizes an iron metalloporphyrin in its free form to form an inactive dimer. In this process, two metalloporphyrin catalysts are consumed. This inactivation pathway occurs predominantly in the dissolution of the catalyst in a liquid medium, where there will tend to be a very high transient concentration of metalloporphyrins, and it is important to keep other vital elements for

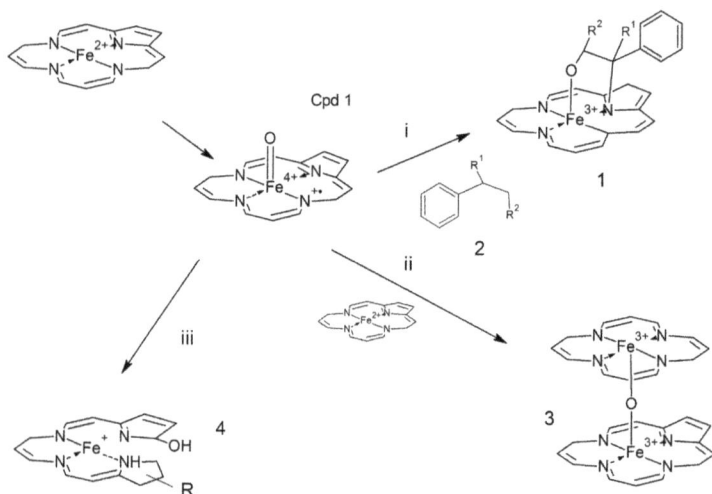

Figure 4. Main reaction pathways of metalloporphyrin degradation in the catalytic cycle of ligninases. Path i occurs with the proximity of a free radical in an alkyl chain to a pyrrolic nitrogen making a covalent bond with the heteroatom, in a process known as N-suicidal destruction, an irreversible process. Pathway ii involves a mutual attack of two metalloporphyrins, one in the form of Cpd I oxo-ferryl to form a stable, non-reactive dimer. Path iii represents multiple oxidative reactions of Cpd 1 or Cpd 2 in the meso or β-pyrrolic positions of the porphyrin macrocycle, which by rearrangement, break the ligand, with destruction of the aromatic character and even the opening of the ring.

the catalytic cycle absent, namely the oxygen donor, in order to reduce the probability occurrence of this degradation.

The oxo-ferryl species in the central metal can also access the β-pyrrolic positions (positions a and b in Figure 3) and meso porphyrin (m in Figure 3) of the porphyrinic macrocycle; which can lead to the intramolecular oxidation of these positions, resulting in the insertion of oxygen in these positions. This can lead to subsequent reactions that result in the resonance of the porphyrin aromatic ring (path iii) being interrupted and, eventually, ring opening. This pathway is an example of autocatalytic oxidative destruction, occurring very quickly, in a few seconds, when adding hydrogen peroxide to protoporphyrin IX solutions, leading to its discoloration.

The vulnerability of metalloporphyrin to the catalytic conditions of delignification reaction makes it important to select suitable structural characteristics in the ligand capable of minimizing the occurrence of degradative steps. The structural characteristics have been known over the years of research and supported by the parallel development of new methods of porphyrin synthesis, generally structures that are difficult to obtain holding more elaborate structural details. The appropriate structural features require porphyrins with a modified environment around the central metal and this can be done by introducing groups or atoms substituting in positions susceptible to oxidative attack—meso or β-pyrrolic porphyrins—those closest to the central metal. One possibility is to add phenyl groups to the meso positions, in a family of porphyrins called meso tetrakis-phenyl porphyrins (structures 2 in Figure 3). These will form an open cage-like structure around the central metal, obstructing the porphyrin's susceptible positions to access the central metal intermediates (Cpd 1 and Cpd 2) and preventing destruction of the catalyst. With metalloporphyrins having these characteristics, more stable biomimetic enzymatic models were built, when compared with the use of protoporphyrin IX or meso alkyl porphyrins.

But even in this structure, the radical intermediates formed in the central metal are reactive enough to destroy the ligands through mechanisms of intramolecular reaction (Oszajca et al. 2016). Thus, a simple phenyl group in the meso positions does not solve the problem of ligand bleaching in biomimetic systems. Some specific reaction conditions are especially demanding for the catalyst, such as the low substrate concentration or the difficult access to the lignin phenylpropane positions. In these conditions, the degradation steps mentioned above become predominant and again lead to the degradation of the biocatalyst (Banfi et al. 1988).

For many years, only porphyrins such as meso tetrakis-phenyl porphyrin (TPP) could be synthesized relatively easily, a fact that led to the biomimetic models for both cytochrome P-450 and peroxidases being essentially built with their metal complexes (Gonsalves and Pereira 1985). These, although they have an environment around the central metal that is much more stereo impeded in relation to protoporphyrin IX, show to have weak to moderate stability in relation to the degradation pathways. Other porphyrins with different structural characteristics were needed to increase the stability of the ligand and potentially build more efficient models. Different structural adaptations have been tried, based on the family of phenyl-porphyrins, but with the addition of atoms or groups that could improve the isolation of the central metal and oxo intermediates. The incorporation of bulky substituents in both ortho positions of the phenyl ring prevents the oxo intermediates from approaching the positions susceptible to oxidation. The main problem to be solved was to find methods of preparing these porphyrins with satisfactory yield.

Several large groups in the meso porphyrin positions have been studied, such as anthranilic, mesityl, among others. These porphyrins could be prepared in low yield by the method of Adler (Adler et al. 1964) and their use tested in biomimetic delignification systems. Its use avoids the formation of complex molecular aggregates as a formation of the μ-oxo dimer (Traylor et al. 1984). However, even with these bulky groups, made with organic molecules that are still labile compared to the demanding reaction intermediates, they also lead to the destruction of porphyrin. This is because the chosen meso substituted groups have C=CH or CH hydrogens, and these are capable of reducing the oxo-ferryl complexes. The search for other characteristics was necessary.

A step forward was the use of atoms of equivalent size instead of organic groups as substituents in the ortho positions of the phenyl groups. At the same time, it was observed experimentally that the introduction of atoms like F or Cl at β-pyrrolic positions increased the stability of metalloporphyrin to degradation. The discovery that the use of electronegative atoms both at the level of the meso phenyl group and in the β-pyrrolic position gave greater stability to these models opened a new and different approach to solve the problem (Traylor et al. 1984). An example of an initial observation of this effect is the great stability of the metal complex of pentafluorophenyl porphyrin in the oxidation of alkanes, with greater stability to degradation compared to TPP. Electronegative atoms seem to stabilize oxidative intermediates, favoring aromatic ring resonance forms of the oxo complex, which remove electrons from the ferryl radical species (Xie et al. 2018). A comparison of oxidative degradation in a set of metalloporphyrins that have one or both of the characteristics of the aspects mentioned above, allows to conclude that only those that present as two conditions simultaneously— bulky and electronegative substituting atoms—are strongly active and stable in biomimetic oxidation. A study of a range of porphyrin ligands with these properties found one of the most desirable metalloporphyrins among the listed structures which are the iron complexes of meso tetrakis (2,6-dichlorophenyl) porphyrin (TDCPP, see Figure 3). This complex has characteristics of robustness and versatility very close to the ideals, allowing the construction of catalysts with activity close to natural peroxidases. Its octahalogenated β-pyrrolic derivatives, with fluorine as the preferred atom, would be very interesting to see, but it is not clear whether they have already been synthesized—for example, perfluorophenyl porphyrin or "Teflon" porphyrin.

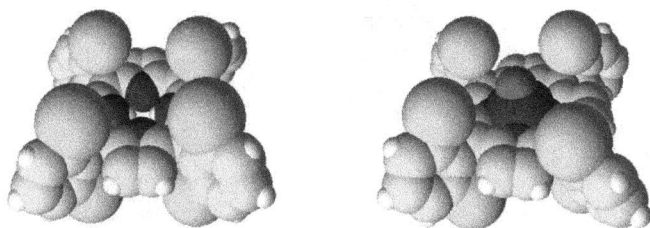

Figure 5. 3D structure of TDCPP (left) and its oxo-ferryl complex (right). The four chlorine atoms form a cage-like structure around the oxygen, preventing it from approaching the meso or β-pyrrolic positions of the macrocyclic, preventing the self-destruction of the catalyst.

The reasons for such porphyrin ligands being very stable to the oxo intermediate is the difficulty in oxidizing bulky and electronegative atoms that surround the oxo ferryl intermediates (Cpd 1 and Cpd 2), while allowing the contact of the substrate with the high valence oxo complexes. Figure 5 shows a 3D molecular model of TDCPP (on the left, Figure 5) and the corresponding oxo-ferryl complex (on the right, Figure 5). The position of the four chlorine atoms can easily be seen, preventing the access of the central ferryl oxygen to the external positions of the macrocycle.

TDCPP is difficult to synthesize, however, methods have been developed that allowed its preparation with better yield. In order to be used in biomimetic delignification models, it is necessary to enter the highly hydrophilic hemicellulose/lignin domain. TDCPP is a very hydrophobic ligand—metal complexes are more soluble in water, but to facilitate their entry into the biomass domain it is necessary to make it more soluble in water. This can be achieved in high yield by chlorosulfonation of the 2,6 dichlorophenyl groups, followed by hydrolysis to sulfate. With the TDCPP this process allows the preparation of the tetra sulfate derivative (TDCPPS) in high yield (Gonsalves et al. 1996).

Few or no attempts are found in the literature to the use of biomimetic models of biomass delignification using this very robust porphyrin described here—Fe (TDCPPS). One reason is the difficulty of synthesis and low global yield, the unavailability of these compounds in the market and, if so, at a very high price. For this reason, the results described in the literature are obtained with less stable water-soluble porphyrins (for example, TPPS), but existing in the market (Ma et al. 2018). The ineffectiveness of using these metalloporphyrins when applied to delignify wood, leads to the coupling of other oxidative chemical methods to assist the process (Yao et al. 2018) with promising results that show the feasibility of using synthetic "enzymes" to effectively carry out the complete delignification of lignocellulosic materials.

3.2 *Immobilization of metalloporphyrin*

When possible, mainly in the cytochrome P450 models as in the case of hydrocarbon oxidation, the immobilization of the synthetic catalyst on a solid support simulates the cage found in natural enzymes and increases its stability and makes its reuse practical (Hu et al. 2017, Hu et al. 2018). The immobilization of metalloporphyrin in a support solid is an imperfect approximation to the isolation conditions that the heme group has in the hydrophobic cavity of ligninases. The consequence is to make an expensive catalyst cheaper by increasing its reusability.

In the present objective—delignification—the catalyst must enter the hemicellulose/lignin domain and immobilization on extensive supports is not feasible, as it is required that the catalyst enter an already highly reticulated matrix. However, small size immobilization supports, which would not prevent their entry, can be considered, if possible by increasing catalytic activity such as groups with possible axial ligands. This may allow the use of porphyrin that is less stable but still more appropriate for performing the catalytic delignification cycle. Other examples could be the connection of sugars to make the molecules more soluble and thus facilitate their entry into the hydrophilic media.

3.3 Preparation of porphyrins

Porphyrins are naturally occurring compounds in many enzymes involved in several important biological metabolic processes, such as a prosthetic group, as oxygen/carbon dioxide transporters, cytochromes, peroxidases, catalase, among others. Synthetic counterparts have a wide range of use in different applications, such as notable examples in photodynamic therapy (PDT), with use in cancer detection and therapy (Deng et al. 2020), in the preparation of blood substitutes (Kitagishi et al. 2017), in catalysis for adding oxygen to a variety of substrates (Meireles et al. 2020), in the design of new molecular electronic devices (Ahmed and Manna 2020), among many others. Meso tetrakis phenyl porphyrins have almost exclusively been used in different applications due to their ease of synthesis and isolation. The optimization of its effectiveness in such applications required the design and preparation of porphyrins with the specific requirements to fulfill the desired properties, such as spectral modifications, solubility manipulation, etc. Wider use and manipulation of their properties is necessary, so the evolution of synthetic methods that can make them easier to prepare, with lower prices is desirable.

The classic general method for the synthesis of meso tetrakis aryl or alkyl porphyrins was developed almost a century ago through the Rothemund reaction. The method allows the preparation of some types of porphyrins, such as meso tetrakis phenyl porphyrin (TPP), and uses as building blocks the pyrrole and aldehydes or acetal derivatives, in a single step, under condensation conditions that allow the cyclization of monomers and oxidation to porphyrin usually in a sealed tube at temperatures as high as 220°C for 48 hours (Rothemund 1935, Rothemund and Menoti 1941). For asymmetric porphyrin, the previous synthesis of dipyrrilmethanes can be subjected to subsequent cyclization in tetrapyrrilmethane to produce porphyrinogen or partially oxidized derivatives, and by total oxidation they can be converted into porphyrins (Liu et al. 2019). These processes are time-consuming and difficult to optimize, generating porphyrin in very low yield.

The cyclization to porphyrin needs the correct positioning of the tetramer and, statistically, it has a low chance of occurring, constituting the main product of the reaction of polymeric solids of polydipyrrilmethenes. This method has been improved over the years. The latest versions of the process make porphyrin synthesis possible in cases where it was not possible. A first step is to carry out a condensation reaction of pyrrole with an aldehyde under reflux of acetic or propionic acid, in partially aerobic conditions. This constitutes the method of Adler (Adler et al. 1964), a method in a single step, from which some porphyrins crystallize directly from the reaction medium with a reasonable yield. The most favorable cases are a meso tetrakis phenyl porphyrin (TPP), where its yield can be reached up to 30%. Meso-tetra alkyl porphyrins remain inaccessible through this process. Subsequently, a two-stage synthesis for the preparation of meso-alkyl porphyrins was developed, based on the analysis of the mechanism of synthesis of porphyrins by Gonsalves (Gonsalves and Pereira 1985). The steps involve the initial formation of porphyrinogen by condensation of the appropriate pyrrole with aldehydes or acetals in a non-oxidative medium, at low temperature, under anaerobic conditions, and acid catalysis (see 3 in Figure 6) which is subsequently subjected to oxidation with separate oxidant added to the reaction medium at a higher temperature to carry out the conversion of porphyrinogen to porphyrin (products 4 to 6 in Figure 6, see below). Based on this method, Gonsalves achieves a synthesis of meso tetra alkyl porphyrins inaccessible by all known conventional methods.

3.3.1 *Intermediates and by-products in the preparation of porphyrins*

The visible absorption spectra observed in the reaction coordinate that turns pyrrole and aldehydes into porphyrins can give a more accurate idea of the intermediates involved in the reaction. For reactions carried out in a single vessel step, such as Rothemund conditions, the spectra obtained are difficult to analyze and poorly resolved due to the presence of a large number of absorbent species of the polymeric type, where only one band stands out— the Soret characteristic of the porphyrins with their high molar absorptivity value, which can reach values of around 500,000.

In two-step methods, the initial preparation of porphyrinogen followed by its subsequent controlled oxidation can, however, reveal the intermediates involved in the process. Tetrakis-4-methoxyphenyl porphyrinogen solutions in anaerobic conditions in the absence of any oxidizer do not have any band in the visible spectrum. The initial intermediates must involve the oxidation of one (or two) porphyrinogen methane positions to result in porphodimethenes (4 in Figure 6). The oxidation of a third position of methane gives rise to isoporphyrin (5) which quickly isomerizes to porphyrin. Careful spectral observation of the slow addition of oxidant to tetrakis-4-methoxyphenyl porphyrinogen solutions shows three bands, one the characteristic of

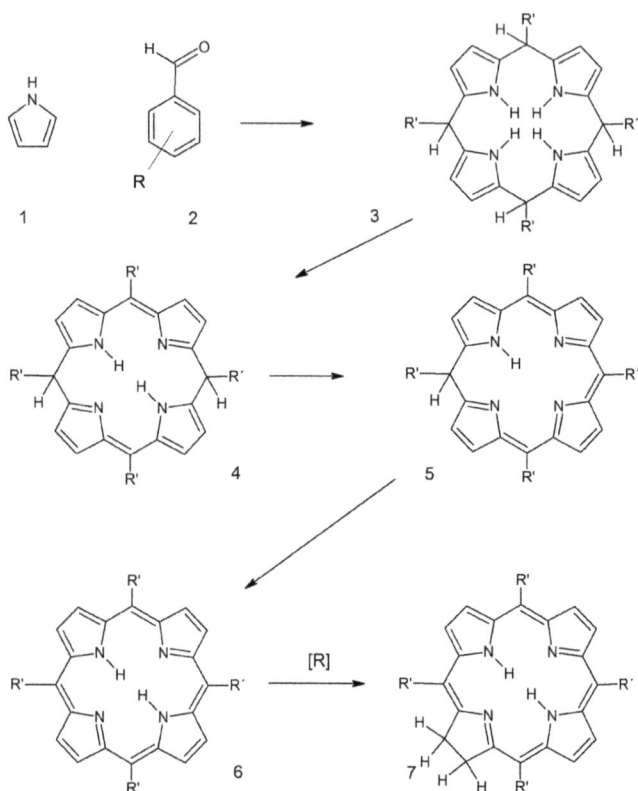

Figure 6. Step-by-step synthesis of porphyrin from pyrrole (1) and an alkyl or aryl aldehyde (2) via cyclization in porphyrinogen (3). The double oxidation of this produces porphodimethenes (4), which by oxidation generate isoporphyrin (5). It quickly isomerizes porphyrin (6). Some synthesis reactions generate chlorin (7) as a by-product, probably obtained by reducing porphyrin.

porphyrin, another the characteristic of porphodimethenes and a third unassigned band. This is observable at 465 nm and cannot be the porphyrin dication (Porf H2 2+), whose characteristic band is only observable in an acidic environment. Once the spectra are recorded in an alkaline medium, another intermediary in the multi-step formation of porphyrins will be observed. The wavelength of the band, 465 nm, between 495 nm (porphodimethenes) and 424 nm (porphyrin), suggests an intermediate chromophore between the two preceding substances, the possibilities being isoporphyrin or phlorin. This scheme adjusts to the number of absorption bands observed in the oxidation reaction coordinate of porphyrinogen to porphyrin.

The intermediate phlorin is sometimes referred to, but never confirmed in porphyrin synthesis. In the direct synthesis of meso tetrakis aryl porphyrins it was never isolated. Phlorin is isomeric from porphodimethenes, so its formation should be independent of the presence of oxidants. In the reduction of uroporphyrin and coproporphyrin, a phlorin was assigned as an intermediate (Gonsalves and Pereira 1985). Its absorption spectrum presents a main absorption characteristic at 440 nm and a small broadband at around 730–740 nm (O'Brien et al. 2019). Studies carried out on the oxidation of meso tetrakisaryl porphyrinogens do not report spectroscopic evidence of this intermediate. The band observed in the porphyrinogen oxidation pathway may also correspond to isoporphyrin (Mwakwari et al. 2011). Isoporphyrins are very difficult to observe in the preparation of porphyrins (Hong and Smith 1992, Xie et al. 2002). Recently, isoporphyrins have been able to be synthesized and their spectra recorded, due to their instability often in the form of metal complexes (Singh et al. 2020). The visible absorption spectra described for isoporphyrins refer to a main band at about 460–470 nm and the "Q" bands in the region of the near-infrared spectrum, at about 800–950 nm, with relative high intensity of absorption in that region. This makes these compounds interesting in techniques such as Dynamic Phototherapy (PDT). The third band observed in controlled oxidation of porphyrinogen (see Figure 6) may, therefore, be the transient presence of isoporphyrin.

Another related structure often found as a by-product in porphyrin synthesis is Chlorin (7 in Figure 6), which is a reduced form of porphyrin (Serra and Gonsalves 2010). The origin of chlorin in the preparation of porphyrins remains controversial. Chlorin may be formed by reducing porphyrin, since there are strongly reducing species in the reaction conditions of porphyrin synthesis, especially polymers of solid dipyrrilmethane, the main product of the reaction. Chlorin contamination with porphyrin synthesis can be eliminated by conversion to porphyrin by treatment with oxidizing agents.

The comparison of the two methods of preparing porphyrins—in one or two stages—indicates that the first is simple and convenient, sometimes giving low or zero yields; the two-stage synthesis is laborious, requires control of precise experimental conditions, and is more prolonged and laborious. In this, yields in a specific range of porphyrins are higher (Lindsey et al. 2010). It would be more convenient if porphyrins could be synthesized in high yield by the first procedure, simpler and in a single step.

3.3.2 *Improved methods for preparing porphyrins*

An improvement of the Adler method was achieved, which greatly increased the synthesis resources in a single step, the synthesis and the easy isolation of many meso tetrakis aryl porphyrins that did not have a preparation method (Gonsalves et al. 1991a). Most importantly, a very wide range of porphyrins for which there was no known method of synthesis, allowed for its preparation, for the benefit of several fields of application from cancer treatment to electronic hardware devices (Xia et al. 2019). One of the porphyrins made available by using this method was TDCPP, a convenient porphyrin for the preparation of systems capable of mimicking the activity of ligninases. This new method arose from the observation of unexpectedly high yields in the preparation of mixed porphyrins related to TDCPP but with a nitrophenyl group, carried out by the Adler method. As an example, when preparing by that method meso 1-(4-nitrophenyl), 5, 10, 15-(2,6-dichlorophenyl) porphyrin, in which the use of a mixture of both aldehydes is necessary to obtain a statistic mixture of meso-substituted porphyrins, the yield obtained in the preparation of TDCPP, a by-product in this particular synthesis, is greater than that obtained when preparing only TDCPP (Varejão 1989). A careful analysis of the differences in the reaction media clearly points to the involvement of the presence of nitro groups in the medium of porphyrin synthesis with mixed meso groups. This led to the addition of nitrobenzene to the Adler reaction medium, an experimental detail that proved to be sufficient to greatly improve the efficiency of the method in a single step in the preparation of a large number of porphyrins. Thus, the main difference between this most recent version and the previous one by Adler is the crucial presence of nitrobenzene in the reaction medium. Porphyrins, prepared at reflux in mixtures of acetic or propionic acid and nitrobenzene (70/30, v/v), result in yields superior to the classic method (Goncalves et al. 1991), usually multiplied by about two or three times and reaching, in the case of meso tetrakis (4-methoxyphenyl) porphyrin, one of the highest known yields in the preparation of porphyrins, equal to 78%. For some of the classically difficult syntheses, such as with meso tetrakis (2,6-dichlorophenyl) porphyrin (TDCPP), the increase in yield is about seven times, going from 0.7% (Kim et al. 1972) to 5.0 % (Gonsalves et al. 1991b). This method has been used over the past few years to successfully prepare a wide range of new porphyrinic structures ((Pinto et al 2019, Xia et al. 2020, Kitagishi et al. 2017). Examples are ethinyl-bridged porphyrin (Xu et al. 2015) and meso-tetrapyrazolyl porphyrins (Guo et al. 2005), among many other examples. An additional advantage of using nitrobenzene as a reaction component is the easy isolation of porphyrin, which is usually done by simple filtration, as porphyrin crystallizes directly from the reaction medium or after the addition of small amounts of methanol. The reactions that have been made with the presence of nitroarenes can, in some cases, generate considerable amounts of chlorin which, in the present case, is not a problem for the development of the catalyst. The presence of high chlorin content has not yet been explained (Serra and Gonsalves 2010). The exact effects of the presence of nitroarenes in increasing the yield in the synthesis of porphyrins have not yet been fully characterized. Apparently, nitrobenzene is not reduced, with only traces of its reduction products, nitrosobenzene and aniline being detected (Cristiano et al. 2003). The formation of H_2O_2 in the course of the reaction was monitored and a higher concentration is detected in the presence of nitrobenzene, which seems to indicate the possibility of its involvement in the effects of improving synthesis, since hydrogen peroxide can assist the oxidation of intermediates to porphyrin (Doctoral Thesis, Varejão 2001). Nitrobenzene can serve as a free radical carrier, facilitating the formation of reactive oxygen species (ROS), thus explaining the presence of hydrogen peroxide in greater concentration in the reaction medium.

Another possibility may be related to the characteristics of black solid materials, polymers of polypyrrilmethenes that constitute the main reaction product. New materials with strong absorption in the infrared area have been synthesized based on dipyrrilmethane derivatives (Zhang and Jin 2018). The reaction temperature normally made in porphyrin synthesis is the boiling point of the used organic acid, acetic or propionic acid, in the range of 120–140°C. The emission of infrared radiation at these temperatures may be sufficient to excite polypyrrilmethenes and induce the formation of the h+/e– pair, both strongly oxidizing and reducing species, respectively. h+, by contact on the entire solid surface with acetic acid or nitrobenzene may generate HO· radicals (Silvestri et al. 2019) that may be the species detected in the hydrogen peroxide detection tests, and the species with a great potential oxidant capable of carrying out oxidation of porphyrinogen. The generation of electrons, capable of reducing the porphyrin already formed, result in the large concentration of chlorin observed, thus explaining its origin. Therefore, the origin of chlorin will not come from the isomerization of porphodimethenes, which would involve a complex movement of both hydrogen ions and double bonds. This scheme can also explain observations made over the years, concerning the improvement of porphyrin synthesis in a single step, which in many cases is improved by the addition of metallic salts to the reaction medium, such as $ZnCl2$ (Banala et al. 2014).

A subsequent adaptation of the two-step method disclosed by Gonsalves for the mesoalkyl porphyrin series (Gonsalves and Pereira 1985) was later adapted by Lindsey (Lindsey et al. 2010) for the meso aryl series. The results proved to be very good in the preparation of a large number of porphyrins and it is a valuable method in the preparation of some types of ligands as well.

The development of new methodologies that allow the synthesis of porphyrins with new functional characteristics facilitates the construction of better enzymatic models. Regarding porphyrins with the potential to be used in the preparation of efficient delignification catalysts, Table 1 shows some

Table 1. Yield of synthesis of porphyrin ligands with potential use in the preparation of biomimetic delignification catalyst.

Porphyrin ligand[1]	Method		
	Acetic/propionic acid[2]	Acetic acid/ nitrobenzene[3]	Two step method[4]
	Yield (%)		
2,6-dichloro	0.7	5.0	24.3
2-chlorophenyl	3.3	20.0	28.0
pentafluoro	10.6	29.0	24.3
2,4,6 trimethyl	4.3	6.0	29.7

1. Meso ligand in porphyrin, 2. Adler method (Adler et al. 1964), 3. Gonsalves et al. 1991a, 4. Lindsey et al. 2010.

examples of porphyrin ligands that can already be prepared by the various methodologies discussed above.

4. Preparation of a robust delignification catalyst

To prepare a robust biomimetic catalyst for lignin peroxidases, the use of TDCPP is the best choice, among the structures with possible synthesis, and must be subjected to chemical modification for the tetra sulfate derivative to guarantee its solubility in water and compatibility with the hemicellulose/lignin domain. The final conversion to the metallic iron (II) complex gives rise to the biomimetic catalyst. Methods that allows its preparation are described below.

4.1 Preparation of 5,10,15,20-tetrakis (2,6-dichloro-3-sulfatophenyl) porphyrin (TDCPPS)

To prepare the TDCPP, any of the methods described below can be used, the first is more practical and fast and produces porphyrin in a yield of about 5.5%, the second is more laborious and expensive, producing the same ligand with a yield of about 20%.

Method 1: Prepare a 250 mL double-necked flask equipped with a reflux condenser and a rubber septum. In the flask place the necessary aldehyde (20×10^{-3} moles) in a mixture of acetic acid (70 ml) and nitrobenzene (30 ml). The mixture is heated to reflux and then pyrrole (20×10^{-3} moles) is added. The mixture is kept under reflux for one hour and, after this period, the heating is switched off. The mixture should be left for slowly cooling for 14 h and then filtered through a sintered glass funnel. Porphyrin is obtained in the form of violet crystals, which must be washed with methanol (3×10 mL) and dried in an oven at 60°C. If no porphyrin crystallizes, methanol (30–50 mL) should be added to the reaction medium, homogenize the mixture and stored in a refrigerator for 14 h, at temperatures close to 0°C. After this period, the solution is filtered and the porphyrin is collected as a purple crystalline material.

Method 2: To a 3 L round-bottom flask, equipped with a magnetic stirrer, add CH_2Cl_2(2 L), previously passed through an alumina column (degree of activity I—column dimensions –10 cm × 5 cm). The solvent is bubbled with nitrogen for 15 min. for removing oxygen. Under nitrogen, the required aldehyde (1.5×10^{-2} moles) is added followed by distilled pyrrole (1.5×10^{-2} moles). After the solution has been homogenized, BF3. Et2O (45%; 240 μL) is added. The reaction mixture is left at room temperature for 3–4 hours. After this period, triethylamine (300 μl) is added, followed by a solution of hydrogen peroxide, H_2O_2 (8 g of H_2O_2 at 35% in 400 ml of acetic acid). The solution is stirred for 1–2 hours, with the bottle open to the atmosphere.

After this period, the entire solvent is evaporated and the residue dissolved in a small volume of chloroform. Porphyrin is isolated and purified by flash chromatography column elution (15 × 5 cm -alumina activity grade II), using chloroform as the eluent. The colored band of the porphyrin is collected and the solvent evaporated, in yield in the range of 20–25%.

Preparation of 5,10,15,20-tetrakis (2,6-dichloro-3-sulfatophenyl) porphyrin (FeTDCPPS) is made by the following procedure.

A 100 mL flask is equipped with a reflux condenser. TDCPP (200 mg; 2.25×10^{-4} moles) and chlorosulfonic acid (12 mL, 2.1×10^{-3} moles) were placed in the flask. The mixture was stirred and heated in an oil bath at 90–100°C for 2 h. After this period, the reaction mixture was cooled and ice was added to destroy the excess chlorosulfonic acid. For the reaction, the medium was added $CHCl_3$ (30 mL) and the organic layer separated. Purification of porphyrin by chromatography on silica gel (G-60; 20 cm × 3 cm), using $CHCl_3$ as eluent. The first band, corresponding to the TDCPP that did not react, was put aside. The second band gave 280 mg of required porphyrin ($\approx$ 93% yield). The chlorosulfonic group must be hydrolyzed to obtain the corresponding sulfate derivative. To make this transformation in a 100 mL round-bottom flask, equipped with a condenser, add 5,10,15,20-tetrakis (2,6-dichloro-3-chlorosulfonylphenyl) porphyrin (200 mg) and distilled water (10 mL). The mixture must be refluxed for 24 h. Chilled, acetone (30 ml) is added to give a precipitate of approximately 88 mg of crystalline porphyrin sulfonic acid (85% yield).

4.2 Metalation of porphyrin

The final delignification catalyst includes a Fe (II) atom in its central position. The insertion of metal or metallization is done by reaction with salts of the required metal, in the present case with iron (II) salts. Metallization reactions must be carried out under a nitrogen atmosphere to avoid oxidation of iron (II) to iron (III), since the latter has difficulty in forming the corresponding complex. In a 100 mL flask equipped with a condenser, place the necessary porphyrin (for example, 4–5 × 10^{-5} moles), and an excess amount of metal salt to favor metallization, such as twice the molar amount necessary, for example, iron (II) acetate (1.0×10^{-4} moles), and distilled water (20 mL). The whole is refluxed for 24–72 h. The end of the reaction is determined by visible spectroscopy, observing the disappearance of the Q bands, characteristic of the porphyrin-free base. The water is evaporated under reduced pressure and the residue dissolved in distilled water (2 ml). This solution is passed through a Sephadex G 15 column (25 × 3.5 cm) using water as an eluent. The second band is collected; corresponds to the desired metalloporphyrin.

Finally, the water is evaporated from the eluate as before, and the metalloporphyrin was dissolved in a minimal amount of methanol and then precipitated by the addition of acetone.

4.3 *Construction of an experimental procedure for biomass delignification*

The conditions for constructing a biomass delignification procedure must meet certain conditions to respect some limitations described previously in the mechanism of ligninases. The first is to avoid excessive concentration of the oxygen donor cofactor, hydrogen peroxide either in biomass or in the aqueous medium as a precaution for the formation of non-active Cpd 3 (see the reaction mechanism above). This recommends a slow, gradual addition of hydrogen peroxide to the delignification process. The use of a programmable syringe pump is appropriate, with addition times up to 45 min. The concentration of the metalloporphyrin catalyst should also not reach high values in the presence of oxidants to prevent its immediate inactivation through the formation of μ-oxo dimer. The most likely situation in which this can happen is in the process of dissolving it in an aqueous medium, and it is recommended that it be done in the absence of any oxygen species. The biomass to be used must have the hemicellulose/lignin domain fully hydrated, which means that from dry biomass a period of humidification must be foreseen. Only in this form will it be possible for the catalyst to diffuse into the biomass, the subsequent addition of hydrogen peroxide is adequate. Thus, a typical biomass de-lignification model uses 1.0 g of dry biomass (105°C) ground (< 500 μm) moistened in 50 mL of water in a 200 mL flask, with a magnetic stirrer. The sample is hydrated for at least four hours with stirring. 50 mg of Fe (II) (TDCPPS), final concentration ≈ 1.5×10^{-6} M, are dissolved in 50 mL of distilled deionized water and the solution added to the biomass flask (final volume of 100 mL). To it 1 μL of pyridine (final concentration of 12 μM) is added to function as an axial ligand. The system is closed by a rubber septum and heated to the desired temperature, usually under 60–100°C. When the temperature stabilizes, a syringe pump is adjusted to add H_2O_2 (30%) at a rate of 0.138 ml/min for one hour or another desired time (total volume equal to 8.3 ml). The reaction is left under these conditions, under stirring for 24 hours. After this period, the mixture is filtered through a porous crucible filter plate, previously heated to 130°C, cooled in a desiccator and weighed. The biomass is washed three times with hot water until a transparent filtrate is obtained and transferred to a drying oven, at 105°C for 6 hours. After this period the crucible is cooled again in a desiccator and weighed. The mass loss is calculated as a percentage of the initial mass.

Table 2 shows a set of delignification tests for *Pinus pinaster* and *Eucaliptus globulus* wood under these conditions.

In order to better understand the degree of delignification achieved by using the biomimetic system, the evaluation of the lignin and cellulose content was carried out using the van Soest method, for some of the tests. The values obtained through this method are approximate and some care must be taken in the rationalization of its results. The values found for the treatments referenced in Table 2 are shown in Table 3, with the same numbering.

Table 2. Wood delignification experiments of *Pinus Pinaster* and *Eucaliptus Globulus* using biomimetic metalloporphyrin catalyst, the oxygen donor was hydrogen peroxide under a set of different experimental conditions.

		Biomass[1]	T (°C)	Reaction time (h)	Weight loss (%)	Xylan + others removal[2] (%)	Catalyst concentration (µM)	pH	Severity[4] Ro	Axial ligand	Notes
1	Blank	*Pinus pinaster*	100	2	< 3.0		1.5	–	100.0	–	
2	Softwood	*Pinus pinaster*	100	2	40.0	≈ 20%	1.5	–	100.0	No	1 mm size particle
3	Softwood	*Pinus pinaster*	100	2	30.5	10	1.5	–	100.0	12 µM pyridine	1 mm size particle; Highly discolored relativity to 1
4	Hardwood	*Eucaliptus globulus*	20	23	3.9		3.8	–	6.1	No	0.5 mm particle size; H_2O_2 addition time 1 h
5	Hardwood	*Eucaliptus globulus*	20	23	7.9		3.8	–	6.1	12 µM pyridine	0.5 mm particle size; H_2O_2 addition time 1 h
6	Hardwood	*Eucaliptus globulus*	47	23	7.6		3.8	3.8	38.0	12 µM pyridine	0.5 mm particle size; H_2O_2 addition time 1 h
7	Hardwood	*Eucaliptus globulus*	76	26	17.8		3.8	–	306.5	12 µM pyridine	0.5 mm particle size; H_2O_2 addition time 1 h
8	Hardwood	*Eucaliptus globulus*	80	2	24.2	≈ 3	3.8	2	30.9	12 µM pyridine	Buffer: HCl/KCl; H_2O_2 addition time 1 h
9	Hardwood	*Eucaliptus globulus*	80	2	31.5	≈ 11	3.8	3	30.9	12 µM pyridine	H_2O_2 addition time 60 min; Buffer: HCl/Potassium Hydrogen Phthalate (HPP)
10	Hardwood	*Eucaliptus globulus*	80	2	25.9	≈ 4.4	3.8	4	30.9	12 µM pyridine	H_2O_2 addition time 60 min; Buffer: NaOH/HPP; white product

Table 2 Contd. ...

...Table 2 Contd.

	Biomass[1]	T (°C)	Reaction time (h)	Weight loss (%)	Xylan + others removal[2,3] (%)	Catalyst concentration (μM)	pH	Severity[4] Ro	Axial ligand	Notes
11	Hardwood *Eucaliptus globulus*	100	2	19.7		3.8	5	120.0	12 μM pyridine	Buffer: NaOH/HPP; H_2O_2 addition time–1 h. Highly white product
12	Hardwood *Eucaliptus globulus*	100	2	16.1		3.8	6	120.0	12 μM pyridine	Buffer: NaOH/Potassium dihydrogen Phthalate
13	Hardwood *Eucaliptus globulus*	100	1.5	20.0		3.8	–	45.0	No	0.25 mm particle size; H_2O_2 addition time–1 h

1. Biomass previously dried at 105°C; the *Pinus Pinaster* biomass was subjected to previous Soxhlet extraction with hexane to remove the terpene fraction; 2. Typical biomass composition—Cellulose (%), Hemicellulose (%), Lignin (%) respectively, lignin content in bold; *Pinus Pinaster*: 46.0, 25.5, **20.0**; *Eucaliptus globulus*: 54.1, 18.4, **21.5**. 3. Difference from theoretical lignin percentage. 4. As defined by the equation Ro = t. exp (T-100/14.75) (Wey et al. 2020, Yuan et al. 2109, Overend and Chornet 1987).

Table 3. Approximate content of lignin and cellulose in biomass samples from delignification experiments of *Eucaliptus globulus* and *Pinus Pinaster* wood with iron metalloporphyrin/hydrogen peroxide.

Biomass	T (°C)	pH	Cellulose (%)	Lignin (%)
Eucaliptus globulus	47		57	20
	60		59	14
	70	6	61	12
	80		63	9
	100		70	5
Pinus pinaster				
		2	67	8
		3	56	6
	80	4	63	10
		5	58	14
		6	65	12

5. Analysis of results

From the results obtained with the biomimetic system implemented with the conditions described above, a detailed analysis of the different parameters in the delignification of wood samples from *Pinus pinaster* and *Eucaliptus globulus* is presented.

5.1 Catalyst behavior in the delignification of wood samples

The catalyst appears to be very active in deconstructing the lignin fraction of the biomass. However, considering that the catalyst has a strong brown color in its active form and the final washes of the biomass result in almost transparent solutions, the evident conclusion is that under the prescribed conditions the catalyst did not survive the conditions of reaction in most experiments. This effect, despite the loss of the catalyst, gives rise to a light-colored biomass, desirable in most treatments. If the catalyst remains in the biomass without bleaching it can result in colored biomass, and in this case, the option for a less robust porphyrin ligand such as meso-2-chlorophenyl porphyrin, a complex that is easier to prepare and economical and still has the potential to ensure good delignification activity, is recommended.

The presence of the axial ligand proved to be, along with what happens in the natural enzyme, essential in the functionality of biomimetic models. In the pair of experiments 2 and 3 (see Table 1), the presence of pyridine significantly increases the activity of the system in delignification efficiency, and there is less activity for non-lignin components, verifiable by the less mass removal, concomitantly with whiteness obtained in the final product. In the absence of that, the characteristic coloring of the catalyst/lignin is observed.

Color is a qualitative sign of delignification, as lignin is a dark brown material. The same effect can be observed in the pair of experiments 4 and 5 (see Table 2) with *Eucaliptus globulus* wood, where in two similar tests, the simple presence of pyridine led to an increase in weight loss from 3.9 to 7.9%. These results also indicate the diffusion of pyridine through the hemicellulose/lignin domain, alongside the catalyst. When comparing experiments 11 and 13 (Table 2) the result verified above is confirmed, resulting in a very white final product. The composition of the buffer solutions is explicit in Table 2 since its components can eventually function as an axial ligand in the catalyst. In the results discussed here, there appears to have been no interference on their part.

5.2 *Delignification efficiency*

To facilitate the understanding of the results, graphs of the effect of the different variables on the composition of the final biomass were made. Thus, the pH value, temperature, severity of conditions (Ro) were related to the observed weight loss. The determination of the approximate composition of the lignin and cellulose content helps the interpretation. Figure 7 shows the effects of weight loss by varying temperature (a) and (b) and pH (graph c). The biomimetic system increases efficiency at higher temperatures where it reaching its greatest weight loss (see below), with an increase in cellulose purity and the lowest residual lignin value (graph b). This aspect should even go beyond the enzymes since the biomimetic system does not undergo denaturation of the protein part, because it does not have it. All components of the system are thermally stable—metalloporphyrin, the axial ligand, with the exception of peroxide, however its involvement is transitory. The higher temperature also increases the diffusivity of the reagents and can allow better access to the biomimetic system to the lignin portions at more intricate points in the biomass structure.

The mass loss expected for the removal of the corresponding lignin exceeded the average lignin content expected in this type of wood in some tests, suggesting that along with the lignin a fraction of hemicellulose may be removed in the process. From approximate models of the wood structure previously presented (see for example Chapter 1), hemicellulose forms part of a reticulated domain with lignin and the existence of connections between the two components is to be expected. Thus, it is to be expected that when the lignin begins to be hydrolyzed and broken into small pieces, structural support for portions of hemicellulose disappears and it is extracted, in an easier way in the process. The biomass weight loss values obtained suggest the conversion of lignin to oligomers and small molecules of phenolic nature, with the part of hemicellulose loss in the process, especially a fraction with low molecular weight DP and accounted for in this fraction. Eventually, hemicellulose may be partially hydrolyzed, in tests carried out at a lower pH value. Hemicellulose linked to cellulose macrofibrils may be partially in position. Weight loss values could suggest hemicellulose removal levels

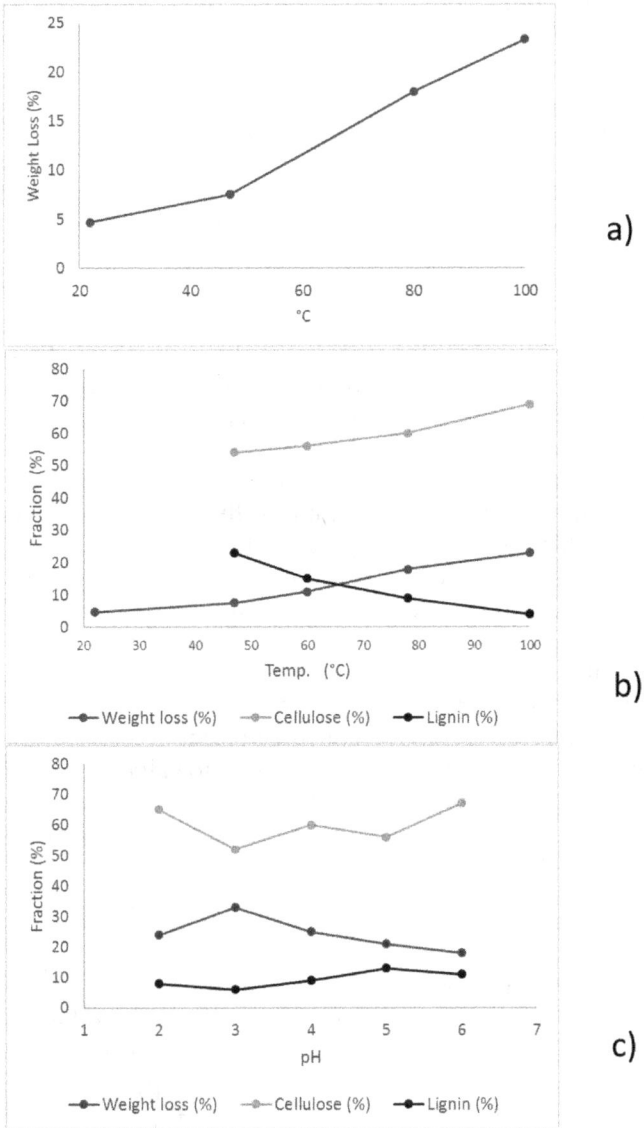

Figure 7. Graphs of the biomass treatment of *Eucaliptus Globulus* with the biomimetic system, showing (a) weight loss versus temperature, (b) cellulose and lignin content versus treatment temperature, (c) the effect of the pH value on the composition of final biomass.

from 10 to 20%. Other components such as minerals and protein can contribute to the weight loss, with the loss of hemicellulose being lesser. The approximate maintenance of the structure of biomass particles suggests the removal of lignin without significant alteration of the three-dimensional structure of hemicellulose (trials numbers 2 and 3 for *Pinus Pinaster* biomass; 8, 9 and 10 for *Eucaliptus globulus* in Table 2). Lignin plus major hemicellulose removal should lead to a main collapse of the cellulose domain producing a compacted material of denser macrofibril arrangements, with small phases of hemicellulose content interface.

5.3 *Effect of temperature*

When analyzing the effect of temperature on the activity of the biomimetic system at temperatures close to the environment, a low catalytic activity is observed, even after 24-hour treatments (tests 4–7 in Table 2). In natural systems, the enzyme is often repressed by low concentrations of the final product, which nevertheless fulfill the requirements of the microorganisms. Human needs try to achieve the greatest results in the shortest possible time. However, at temperatures above about 75°C, most of the lignin begins to be depolymerized at a significantly high rate, steadily increasing up to 100°C, the maximum temperature tested.

In these conditions, time periods as low as 2 h seem to be sufficient to depolymerize most of the lignin present in the biomass. The best result of mass loss in Table 2 corresponding to the test at 80°C, justifying a partial hydrolysis of hemicellulose at the same time. With the determinations of lignin content, biomimetic treatments at temperatures above 80°C present lignin content in the range of 5–7%, considering the precision and error inherent to the evaluation methods used (3–4%)—coupled hydrolysis with gravimetric methods—it can be considered that using treatment at temperatures 80–100°C the lignin content present is residual. The cellulose content increases consistently with temperature to values of up to 70%, which means that the final product may contain some level of hemicellulose. The accounting includes normal fractions of proteins and minerals that remain occluded in the biomass.

5.4 *Effect of pH value*

The effect of pH variation on treatments (see Figure 7) indicates greater loss of mass at pH 3, together with a lower lignin content in the final product. Surprisingly, the cellulose content appears to be lower at this pH value, lower than that found in treatments carried out at pH 2. The hydrolysis of amorphous cellulose through the facilitated access to it by delignification may be operating concurrently for the observed effect. The mass loss of biomass at pH 3 reaches 32% and the lignin content, a minimum at values close to 6% (Tables 2 and 3). Apparently, amorphous cellulose is degraded to some extent at this pH value, to a greater extent than at more acidic values, such as pH 2. The result is not easy to rationalize, but some interference of the acid in the delignification mechanism, probably by allowing access to the amorphous cellulose may occur. The removal of lignin/hemicellulose should allow better acidic access to the amorphous cellulose portions, facilitating the hydrolysis of these cellulose zones with the observed effect.

5.5 *Severity of conditions*

The Severity scale (Ro) defined by Overend and Chornet (1987) is essentially thought of and applied in chemical treatments in the delignification process, and in principle does not apply to enzymatic systems. However, since the present "enzymes" are chemical catalysts in nature, with any characteristic effects occurring from the presence of the protein component, the use of the scale may make sense. The graphs of the Severity variation in the characteristics of the final product are presented in Figure 8, for different delignification tests with *Eucaplitus globulus* wood.

Weight loss when using only time and temperature variables seems to linearly follow the Severity index shown in Figure 8(a), however when considering all the variables tested, a greater dispersion of results is obtained, as can be seen in Figure 8(b). An overall effect of greater mass loss with greater Severity is seen rather as a trend, much less pronounced than using the characteristic Severity variables (t and T) (Wei et al. 2020). This clearly shows the enzymatic-type characteristic of the catalysis which, even with the absence of apoprotein, shows the behavior of the natural catalyst, depending on other factors in the general activity besides t and T.

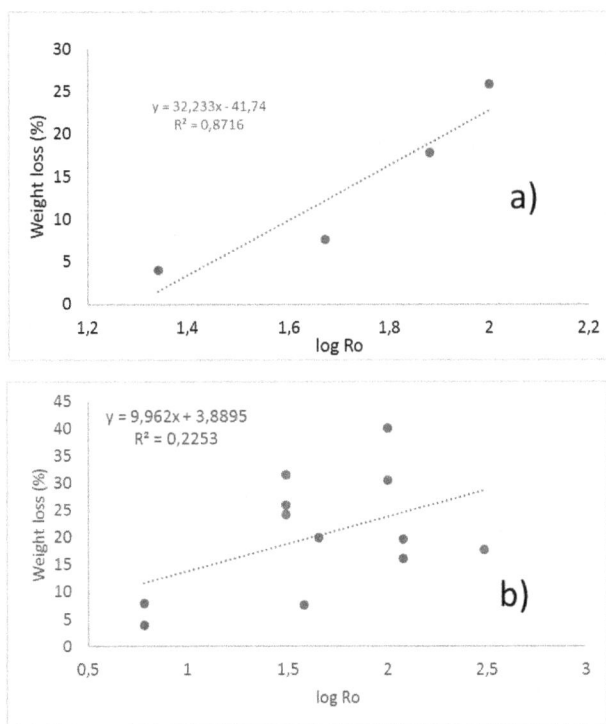

Figure 8. Severity (Ro) graphs versus biomass weight loss expressed as a percentage of *Eucaliptus globulus* biomass; (a) in systems with fixed time and variable temperature; (b) in systems including all variables tested (axial ligand, pH, time and temperature).

5.6 *Particle size effects of biomass*

The particle size was studied in tests 3 and 11 and 13, shown in Table 2. More comparable are tests 11 and 13 since experiment 3 was done with *Pinus Pinaster* biomass. The results obtained in terms of weight loss are similar, suggesting that the degree of granulometry, in the size range in the experimental conditions used in the present works and does not interfere to a large extent in the catalytic delignification process. The preliminary pre-treatment of wetting and the gradual addition of the oxidizer may be the reason.

6. Final notes

The results indicate that in some tests the complete delignification of wood is achieved by the use of a biomimetic system. Contributing to this result are factors such as the use of a highly robust metalloporphyrin, the slow addition of the oxygen donor and the time allowed for the catalyst to diffuse into the biomass.

The availability of a robust porphyrin ligand, such as meso tetrakis -2,6-dichloro, 3-sulfatophenyl porphyrin, allows the construction of systems capable of efficiently removing lignin from biomass in a single vessel process, one step further and desirable when compared to more elaborate processes involving various oxidative steps, such as a secondary Baeyer-Villiger oxidation. Biomass treatment must include conditions to protect the catalyst; avoid the high concentration of metalloporphyrins with oxidants, by gradually add hydrogen peroxide and allow the biomimetic catalyst and cofactors such as the axial ligand to diffuse in the porous hemicellulose/ lignin domain. This mechanism of action will remove the lignin, giving rise to a product with a composition similar to the mixture of phenolic substances, contaminated with some low DP hemicellulose. Phenolic compounds alone have interesting properties that can be used in different fields, from antioxidants to analogs of plant hormones, among many others.

The resulting biomass constitutes a new material–biomass or delignified wood, but maintaining a higher hemicellulose content—a material that has not yet been well studied. It may have stronger mechanical properties, a white color with a characteristic capable of generating a range of new materials, such as the already existing transparent wood, enabling the production of furniture, "glasses" and sustainable materials for other industries.

Another possible use is in the area of carbonization, by any of the existing methods, since in the absence of lignin the formation of turbostratic graphite can occur in a more perfect way. Obtaining and evaluating materials with this structure have begun. It is expected that the graphitization of these materials will give rise to substances derived from biomass, with application in areas such as electronics, batteries, supercapacitors, etc.

A challenge that remains to be faced is the emergence and development of new methods that can make porphyrin ligands more accessible and economical. Fe(TDCPPS) is difficult to find on the market and when available its price is too high. The use of pentafluorophenyl porphyrin or monochlorohydroxyphenyl, the latter of direct synthesis and that by simple metallization produces a relatively water-soluble catalyst, are especially promising in the dissemination of the method. The best study of the various methods of preparing porphyrins with a better understanding of the role of solid polymers in their synthesis, may eventually improve the process and yield of porphyrin making these technologies common and of general use.

7. References

Adler, A. D., Shergali, W. and F. R. Longo. 1964. Mechanistic investigations of porphyrin syntheses. I. Preliminary studies on meso-tetraphenylporphin. Journal of the American Chemical Society 86: 3145–3146.

Ahmed, R. and A. K. Manna. 2020. Molecular-scale engineering of the charge-transfer excited states in non-covalently bound Zn-porphyrin and carbon fullerene-based donor-acceptor complex. Physical Chemistry Chemical Physics 22: 14822–14831.

Banala, S., Wurst, K. and B. Krautler. 2014. Symmetrical tetra-beta-sulfoleno-meso-aryl-porphyrins—synthesis, spectroscopy and structural characterization. Journal of Porphyrins and Phthalocyanines 18: 115–122.

Banfi, S., Montanari, F. and S. Quici. 1988. New manganese Tetrakis (Halogenoaryl) porphyrins featuring sterically hindering electronegative substituents—synthesis of highly stable catalysts in olefin epoxidation. Journal of Organic Chemistry 53: 2863–2866.

Cristiano, M. L. S., Gago, D. J. P., Gonsalves, A. M. D. R., Johnstone, R. A. W., McCarron, M. and Varejao, J. M. T. B. 2003. Investigations into the mechanism of action of nitrobenzene as a mild dehydrogenating agent under acid-catalysed conditions. Organic & Biomolecular Chemistry 1: 565–574.

da Silva, V. S., Meireles, A. M., Martins, D. C. D., Reboucas, J. S., DeFreitas-Silva, G. and Y. Idemori. 2015. Effect of imidazole on biomimetic cyclohexane oxidation by first, second-, and third-generation manganese porphyrins using PhIO and PhI(OAc)(2) as oxidants. Applied Catalysis A-General 491: 17–27.

Deng, J. R., Li, H., Yang, M. Q. and F. S. Wu. 2020. Palladium porphyrin complexes for photodynamic cancer therapy: Effect of porphyrin units and metal. Photochemical & Photobiological Sciences 19: 905–912.

Diaz-Angulo, J., Porras, J., Mueses, M., Torres-Palma, R. A., Hernandez-Ramirez, A. and F. Machuca-Martinez. 2019. Coupling of heterogeneous photocatalysis and photosensitized oxidation for diclofenac degradation: role of the oxidant species. Journal of Photochemistry and Photobiology A-Chemistry 383: 112015.

Fang, Z., Mobley, J. K. and M. S. Meier. 2018. Dramatic simplification of lignin heteronuclear single quantum coherence spectra from ring-and-puck milling followed by oxidation. Energy & Fuels 32: 11632–11638.

Gonsalves, A. M. D. R. and M. M. Pereira. 1985. A new look into the rothemund meso-tetraalkyl and tetraaryl porphyrin synthesis. Journal of Heterocyclic Chemistry 22: 931–933.

Gonsalves, A. M. D. R., Johnstone, R. A. W., Pereira, M. M., Shaw, J. and, A. J. F. D. Sobral. 1991a. Metal-assisted reactions. 22. Synthesis of perhalogenated porphyrins and their use as oxidation catalysts. Tetrahedron Letters 32: 1355–1358.

Gonsalves, A. M. D. R., Varejao, J. M. T. B. and M. M. Pereira. 1991b. Some new aspects related to the synthesis of meso-substituted porphyrins. Journal of Heterocyclic Chemistry 28: 635–640.

Gonsalves, A. M. D. R., Johnstone, R. A. W., Pereira, M. M., de SantAna, A. M. P., Serra, A. C., Sobral, A. J. F. N. and P. A. Stocks. 1996. New procedures for the synthesis and analysis of 5,

10, 15, 20-tetrakis (sulphophenyl) porphyrins and derivatives through chlorosulphonation. Heterocycles 43: 829–838.

Guo, C., Ren, T. G., Song, J. X., Liu, Q., Luo, K., Lin, W. Y. and G. F. Jiang. 2005. Substituted tetrapyrazolylporphyrins: Application in organic light-emitting diodes. Journal of Porphyrins and Phthalocyanines 9: 830–834.

Hilden, K., Makela, M. R., Steffen, K. T., Hofrichter, M., Hatakka, A., Archer, D. B. and T. K. Lundell. 2014. Biochemical and molecular characterization of an atypical manganese peroxidase of the litter-decomposing fungus Agrocybe praecox. Fungal Genetics and Biology 72: 131–136.

Hofrichter, M. 2002. Review: Lignin conversion by manganese peroxidase (MnP). Enzyme and Microbial Technology 30: 454–466.

Hong, X. and K. M. Smith. 1992. Stable isoporphyrin chromophores—synthesis. Tetrahedron Letters 33: 1197–1200.

Hu, G. F., Kang, H. S., Mandal, A. K., Roy, A., Kirmaier, C., Bocian, D. F., Holten, D. and J. S. Lindsey. 2018. Synthesis of arrays containing porphyrin, chlorin, and perylene-imide constituents for panchromatic light-harvesting and charge separation. RSC Advances 8: 23854–23874.

Hu, X. M., Ronne, M. H., Pedersen, S. U., Skrydstrup, T. and K. Daasbjerg. 2017. Enhanced catalytic activity of cobalt porphyrin in CO2 electroreduction upon immobilization on carbon materials. Angewandte Chemie-International Edition 56: 6468–6472.

Ji, X. J., Han, Z. B., Li, J. F., Deng, Y., Han, X., Zhao, J., Zhao, X. M. and C. C. Chen. 2019. MoSx co-catalytic activation of H2O2 by heterogeneous hemin catalyst under visible light irradiation. Journal of Colloid and Interface Science 557: 301–310.

Kim, J. B., Leonard, J. J. and F. R. Longo. 1972. Mechanistic study of synthesis and spectral properties of Meso-Tetraarylporphyrins. Journal of American Chem. Soc. 94: 3968.

Kitagishi, H., Mao, Q. Y., Kitamura, N. and T. Kita. 2017. HemoCD as a totally synthetic artificial oxygen carrier: Improvements in the synthesis and O-2/CO discrimination. Artificial Organs 41: 372–380.

Kudrik, E. V., Afanasiev, P., Alvarez, L. X., Dubourdeaux, P., Clemancey, M., Latour, J. M., Blondin, G., Bouchu, D., Albrieux, F. and S. E. Nefedov. 2012. An N-bridged high-valent diiron-oxo species on a porphyrin platform that can oxidize methane. Nature Chemistry 4: 1024–1029.

Li, B. H., Zhao, M. Y. and F. Zhang. 2020a. Rational design of near-infrared-II organic molecular Dyes for bioimaging and biosensing. ACS Materials Letters 2: 905–917.

Li, R., Wang, X. H., Lin, Q. X., Yue, F. X., Liu, C. F., Wang, X. Y. and J. L. Ren. 2020b. Structural features of lignin fractionated from industrial furfural residue using alkaline cooking technology and its antioxidant performance. Frontiers in Energy Research 8: 83.

Lindsey, J. S. 2010. Synthetic routes to meso-patterned porphyrins. Accounts of Chemical Research 43: 300–311.

Liu, S. Y., Tanaka, H., Nozawa, R., Fukui, N. and H. Shinokubo. 2019. Synthesis of meso-alkyl-substituted Norcorrole-Ni-II complexes and conversion to 5-Oxaporphyrins (2.0.1.0). Chemistry-A European Journal 25: 7618–7622.

Lu, X. J., Zhou, X. F., Zhu, Z. L., Sun, Y. L., Tang, K., Lei, F. H., Liu, Z. G. and T. Wang. 2019. Catalytic oxidation of lignin to aromatics over salen-porphyrin complex as a biomimetic catalyst. Drewno 62: 67–80.

Ma, R. S., Guo, M. and X. Zhang. 2018. Recent advances in oxidative valorization of lignin. Catalysis Today 302: 50–60.

Manno, M., Tolando, R., Ferrara, R., Rezzadore, M. and S. Cazzaro. 1995. Suicidal inactivation of hemoproteins by reductive metabolites of halomethanes—a structure-activity relationship study. Toxicology 100: 175–183.

Mathew, S., Abraham, T. E. and Z. A. Zakaria. 2015. Reactivity of phenolic compounds towards free radicals under *in vitro* conditions. Journal of Food Science and Technology-Mysore 52: 5790–5798.

Meireles, A. M. and D. C. D. Martins. 2020. Classical and green cyclohexane oxidation catalyzed by manganese porphyrins: Ethanol as solvent and axial ligand. Polyhedron 187: 114627.

Mwakwari, S. C., Wang, H. J., Jensen, T. J., Vicente, M. G. H. and K. M. Smith. 2011. Syntheses, properties and cellular studies of metallo-isoporphyrins. Journal of Porphyrins and Phthalocyanin 15: 918–929.

O'Brien, A., McGann, J. P. and R. G. Geier. 2019. Phlorin dipyrromethane + dipyrromethanedicarbinol routes to an electron deficient meso-substituted phlorin with enhanced stability. J. Org. Chem. 25: 7618–7622.

Oszajca, M., Franke, A., Brindell, M., Stochel, G. and R. van Eldik. 2016. Redox cycling in the activation of peroxides by iron porphyrin and manganese complexes. 'Catching' catalytic active intermediates. Coordination Chemistry Reviews 306: 483–509.

Overend, R. P. and E. Chornet. 1987. Fractionation of lignocellulosics by steam-aqueous pretreatments. Philosophical Transactions of the Royal Society of London. Series A, Mathematical and Physical Sciences 32: 1523–536.

Pinto, S. M. A., Vinagreiro, C. S., Tome, V. A., Piccirillo, G., Damas, L. and M. M. Pereira. 2019. Nitrobenzene method: A keystone in meso-substituted halogenated porphyrin synthesis and applications. Journal of Porphyrins And Phthalocyanines 23: 329–346.

Rothemund, P. 1935. Formation of porphyrins from pyrrole and aldehydes. Journal of the American Chemical Society 57: 2010–2011.

Rothemund, P. and A. R. Menotti. 1941. Porphyrin studies. IV. 1 The synthesis of α, β, γ, δ-tetraphenylporphine. Journal of the American Chemical Society 63: 267–270.

Serra, A. C. and A. M. D. R. Gonsalves. 2010. Controlled porphyrinogen oxidation for the selective synthesis of meso-tetraarylchlorins. Tetrahedron Letters 51: 4192–4194.

Silvestri, S., Ferreira, C. D., Oliveira, V., Varejao, J. M. T. B., Labrincha, J. A. and D. M. Tobaldi. 2019. Synthesis of PPy-ZnO composite used as photocatalyst for the degradation of diclofenac under simulated solar irradiation. Journal of Photochemistry and Photobiology A-Chemistry 375: 261–269.

Singh, N. G., Borah, K. D., Majumder, S. and J. Bhuyan. 2020. Ethanol coordinated zinc trimethoxyphenylporphyrin: Structure, theoretical studies and formation of isoporphyrin. Journal of Molecular Structure 1200: 127116.

Traylor, P. S., Dolphin, D. and T. G. Traylor. 1984. Sterically protected hemins with electronegative substituents—efficient catalysts for hydroxylation and epoxidation. Journal of the Chemical Society-Chemical Communications 5: 279–280.

Varejão, J. M. T. B. 1989. Developments in the Synthesis of Meso-tetra Phenyl Porphyrins. M.S. Thesis. Coimbra University, Portugal.

Varejão, J. M. T. B. 2001. Towards the Synthesis of Model Peroxidases. PhD Thesis. University of Liverpool, UK.

Wei, W., Tian, Z. J., Ji, X. X., Wang, Q., Chen, J. C., Zhang, G. C. and L. A. Lucia. 2020. Understanding the effect of severity factor of prehydrolysis on dissolving pulp production using prehydrolysis kraft pulping and elemental chlorine-free bleaching sequence. Bioresources 15: 4323–4336.

Witzler, M., Alzagameem, A., Bergs, M., El Khaldi-Hansen, B., Klein, S. E., Hielscher, D., Kamm, B., Kreyenschmidt, J., Tobiasch, E. and M. Schulze. 2018. Lignin-derived biomaterials for drug release and tissue engineering. Molecules 23: 1885.

Wu, Y., Wu, J., Yang, F., Tang, C. and Q. Huang. 2019. Effect of H2O2 bleaching treatment on the properties of finished transparent wood polymers. Polymers 11: 776.

Xia, Z. J., Yang, H. C., Chen, Z. W., Waldman, R. Z., Zhao, Y. S., Zhang, C., Patel, S. N. and S. B. Darling. 2019. [1,5,6,7] Porphyrin covalent organic framework (POF)-based interface engineering for solar steam generation. Advanced Materials Interfaces 6: 1900254.

Xie, H., Leung, S. H. and K. M. Smith. 2002. Syntheses and some chemistry of stable isoporphyrin systems. Journal of Porphyrins and Phthalocyanines 6: 607–616.

Xie, J. F., Ma, G. F., Ouyang, X. P., Zhao, L. S. and X. Q. Qiu. 2018. Metalloporphyrin as a biomimetic catalyst for the catalytic oxidative degradation of lignin to produce aromatic monomers. Waste and Biomass Valorization 11: 4481–4489.

Xu, Z. J., Mei, Q. B., Hua, Q. F., Tian, R. Q., Weng, J. N., Shi, Y. J. and W. Huang. 2015. Synthesis, characterization, energy transfer and photophysical properties of ethynyl bridge linked porphyrin-naphthalimide pentamer and its metal complexes. Journal of Molecular Structure 1094: 1–8.

Yao, S. G., Mobley, J. K., Ralph, J., Crocker, M., Parkin, S., Selegue, J. P. and M. S. Meier. 2018. Mechanochemical treatment facilitates two-step oxidative depolymerization of kraft lignin. ACS Sustainable Chemistry & Engineering 6: 5990–5998.

Yuan, S. J., Ji, X. X., Ji, H., Tian, Z. J. and J. C. Chen. 2019 An optimum combined severity factor improves both the enzymatic saccharification yield and the functional lignin structure. Cellulose 26: 4731–4742.

Zhang, M. and R. F. Jin. 2018. Rational design of near-infrared dyes based on boron dipyrromethene derivatives for application in organic solar cells. RSC Advances 8: 33659–33665.

Zhang, R., Maltari, R., Guo, M., Kontro, J., Eronen, A. and T. Repo. 2020. Facile synthesis of vanillin from fractionated Kraft lignin. Industrial Crops and Products 145: 112095.

Zucca, P., Rescigno, A., Rinaldi, A. C. and E. Sanjust. 2014. Biomimetic metalloporphines and metalloporphyrins as potential tools for delignification: Molecular mechanisms and application perspectives. Journal of Molecular Catalysis A-Chemical 388: 2–34.

Chapter 4

Bioproducts Derived from Lignin Obtained from Micro- and Nanocrystalline Cellulose Preparation

1. Introduction

From the best knowledge of the structural details of cellulose in biomass, essentially two types of domains are found: the cellulose domain and the hemicellulose/lignin one. The first is cellulose in the form of superarrays of micrometric dimension, with the presence of crystalline parts together with amorphous zones. The variability of the cellulose structure, between crystals and amorphous zones with interlacing of other components, must correspond to configurations that satisfy the local structural needs in the plant. In crystalline cellulose, a strong network of hydrogen bonds is formed at intra- and intermolecular levels, making it one of the known biological materials with stronger mechanical properties. Amorphous cellulose shows a less organized structure with the presence of empty spaces between the macrofibrils, allowing intertwining with both hemicellulose and lignin. This phase is more easily accessible to small molecules and enzymes, eventually allowing them to decompose into monomers. The use of enzymes requires partial removal of the hemicellulose/lignin domain, otherwise access is problematic, and they tend to be adsorbed by lignin.

2. Microcrystalline cellulose (MCC)

Lignin and hemicellulose are removed, and if the regions of amorphous cellulose are decomposed by chemical or enzymatic hydrolysis, a residue is obtained, which is mainly composed of crystalline cellulose.

The resulting particle size occurs in the range of 50–200 μm and this fraction is commonly referred to as microcrystalline cellulose (MCC). MCC has properties that make it interesting for use in different areas, because it is a biocompatible material with good stability to degradation, difficult to attack either by acids or bases, has good adsorption properties, good compressibility, and good transportability from the point of view of movement, making it interesting as an excipient for the pharmaceutical field in industrial tablet preparation (Anwarab et al. 2014). If it is ingested by humans, it constitutes dietary fiber. If the particle size is adequate, it has the ability to act as a thickening and gelling agent, and can function as a substitute for products derived from starch with similar properties, but without any caloric content, which is currently relevant in the fight against the spread of obesity.

More recently, MCC, due to its excellent mechanical properties (see Table 1), has come to be used in other more specialized areas such as the preparation of biocompatible composite biomaterials, along with the further development that three-dimensional printers are beginning to have. Other materials that are beginning to receive great attention include the preparation of cellulose nanoparticles (CNC) and cellulose nanofibers (CNF), obtained from MCC, materials with the potential to generate new materials, for use in medical areas, batteries, supercapacitors, and various types of composites. For these, a large number of end uses can already be considered (see below).

Table 1. Mechanical properties of crystalline cellulose (MCC) and nanocellulose (CNC).

Cellulose type	Young modulus (GPa)	Breaking tension (GPa)	Reference
MCC	25	–	Eichhorn and Young 2001
CNC	150	7.50	Moon et al. 2013
	110–220	2.0–8.0	Ramezani and Golchinfar 2019
	150	7.5–7.7	Wu et al. 2014, Börjesson and Westman 2015

2.1 *Preparation of paper pulp*

The preparation of the MCC is a well defined and stabilized process that has been known for some time and is based on the pulp of the cellulose industry. Some specific variants that generate a final product of better quality, can be obtained by treatment with cotton with mineral acid (Yavorov et al. 2020). The cellulose pulp industry prepares a large volume annually that meets human needs in a wide range of paper products, packaging, etc. While the use of office paper tends to decrease, other uses have been increasing with the attempt to reduce the use of non-biodegradable plastics, such as bags, cotton buds, etc. The MCC preparation for medical uses, filters for scientific research and general use is expected to continue to grow.

The industrial production of cellulose is mainly carried out by one of two types of process, both effecting the delignification of wood biomass. The Kraft process cooks the biomass in the form of chips with 17–22% sodium hydroxide, in the presence of sodium sulfite, at temperatures of 170°C, for 3–4 hours. The total volume of the reactors can reach high values, such as 300 tons biomass per load, in batch process. Hardwood types such as eucalyptus or others such as pine wood are preferred. The use of alkaline conditions has the advantage of not corroding steel equipment, which is desirable from an industrial point of view. The other process in use is the sulfite process, under acidic conditions. In this, sulfur dioxide (SO2) or sulfuric acid are added with sulfite or metabisulfite, to promote the delignification of biomass. In both techniques, final polishing treatments such as bleaching are done using oxidants, either with chlorine (less desirable) or with peroxides. The products obtained in the general process are the pulp and hemicellulose/lignin hydrolysates—the so-called sulphitic liquors. The latter have a composition that, in summary, consists of minerals, alkaline salts, lignin sulfonates related to phenols and sugars. The delignification conditions used in any of the above processes, lead to rapid removal of hemicelluloses and its hydrolysis at the level of monosaccharides. As the ether bonds of C-O-C lignin are hydrolyzed by OH–, depolymerization occurs by producing a mixture of low and medium mass lignin sulfonates (Kuznetsov et al. 2018). The preparation scale makes it possible to obtain cellulose pulp at very competitive prices, for example, Portugal alone produces more than 1 billion tons of paper pulp annually, thus making this purified biomass that benefits from large-scale production the material to be chosen in the preparation of the MCC. At the same time, the pulp has good color properties and an eventually desirable crystallinity index.

2.2 *Preparation of MCC from cellulose pulp*

The usual MCC manufacturing process was described a long time ago, shortly after World War II with the preparation of Avicel; this same method has been optimized over the years (Yue et al. 2019). The method is applicable to raw materials such as pulp from the cellulose industry or pure cotton and consists of a rapid acid hydrolysis treatment of the amorphous regions of the cellulose. Hydrochloric acid is particularly effective under conditions such as 2M concentration, for 15 min at boiling temperature. Other conditions achieve the same objective, such as HCl 1 M, 85°C, a biomass to acid ratio of 1:10 (w/v), with reaction times of 40 to 120 min (Li et al. 2019). After the treatment, the product is filtered and washed with hot water leaving the MCC in a compact form as a residue, which is then dried and decompressed. The separation of the product into fractions of determined particle size and finishing treatments such as additional bleaching and/or milling are available to serve any end user in the industry, such as stationary phase for chromatographic separations, manufacture of laboratory and industrial filters, ashless papers, etc.

A potential disadvantage of the process of preparing MCC from pulp is the crystallinity index of the initial pulp. If the crystallinity of the pulp is low, the extent of hydrolysis of the pulp in the acid treatment can be high, leading to very low yields in MCC, with most of the pulp mass being converted into sugars. As mentioned before, the use of soft wood is often common in the preparation of paper pulp and it may have a low degree of crystallinity. From the point of view of biorefinery, where the complementarity of processes can be found, this effect can be overcome if alternative sugar conversion routes are available, for example their conversion into ethanol. The fermentation capacity must be carried out by quantifying the presence of furfuraldehydes, strong inhibitors of yeast activity. However, both possibilities are not always present, and if only the MCC preparation can be considered, sugar-rich fractions can be wasted. To avoid this effect, a careful study of the properties of the paper pulp is necessary (Laka and Chernyavskaya 2007).

2.3 Preparation of MCC from straw residues

2.3.1 Alkaline thermal delignification

The process described above is in production in a large number of companies, producing various types of MCC for the market. In view of the vast amount of biomass residues, sometimes accumulated in small spaces and for which no useful destination was found other than burning, the production of MCC from them could constitute a useful destination with an economic value (Ohwoavworhua and Adelakun 2010). Especially the by-products of cereal production, straw, as well as the regular thinning of grasses that naturally grow everywhere, are considered a potential source of raw material for obtaining MCC. The reason is very easy to understand; straws, due to the functions they perform in plants, have high levels of crystalline cellulose. In this way a higher yield in obtaining CMC will be achieved. To make this process competitive, it is important to find advantages that go beyond the scale factor in pulp production.

Wheat and rice are two important cereals in the human food chain, indicating that their straw is a waste of abundant biomass. Corn straw has in its composition about 20% (w/w) starch on dry mass basis, and together with the hemicellulose content constitutes an important source of animal feed, making its conversion less desirable. However, it depends on the context in which it is produced and in the same way, it can be included in the process. Other grasses, often growing in fallow land, have the same apical growth that requires high crystallinity in the cellulose. They are scattered, many being collected by local authorities, and disposed of in landfills. This residual biomass may also be subject to study and included in the process. The preparation of MCC from these raw materials is more difficult and elaborate, as lignin and hemicellulose are still present in them, compared to paper pulp raw materials. The steps involved in the manufacture of cellulose pulp that must be included are the delignification of biomass and removal of hemicelluloses followed by removal of amorphous cellulose. Effective experimental conditions for removing lignin require severe reaction conditions, while removing at the same time hemicellulose. However, as hemicellulose can be a valuable bioproduct, there is every interest in recovering it. The preparation of MCC from straw residues initially involves grinding. If its storage is necessary, it must be dried to prevent the growth of microorganisms. The first fraction to be removed, if provided, will be hemicellulose by treatments such as using hot water or prolonged contact with water at room temperature, followed by extraction with slightly alkaline hot water.

The classic conditions for delignification of biomass consume high values of energy, usually through the use of thermal heating at the boiling point of strong base solutions for up to 36 hours. Typical conditions involve cooking ground biomass with aqueous sodium hydroxide (NaOH) solutions in concentrations in the range of 17–22% for reaction times of 15–20 hours at ambient pressure. Table 2 presents a compilation of hemicellulose removal plus delignification results from ground wheat straw, deduced from the weight loss data.

Table 2. Delignification of wheat straw biomass by thermal heating with sodium hydroxide in different experimental conditions.

Wheat straw conditions	[NaOH] % (w/v)	Time (h)	Temperature °C	App. mass removal (%)	Cellulose yield (%)
Theoretical				41.0–63.0	47–59.0
1	5.0	1		24.40	75.6
2	5.0	6		33.60	66.4
3	17.5	6	100	38.10	61.9
4	17.5	12		56.40	43.6
5	17.5	16		56.60	43.4
6	22.5	6		65.60	34.4

Considering the average values of wheat straw composition, with 37.0% cellulose, 27.0% hemicellulose and 14.0% lignin, and 22.0% for the other components, mainly minerals, proteins, soluble sugars, among others; the weight loss corresponding to the fractions of hemicellulose + lignin should be around 41%. If we consider that in weight loss, a partial hydrolyzable fractions of protein and other components, the value can increase to approximately 63.0% of the weight loss. Experimental conditions with an increase in severity are listed in Table 2 from 1 to 6. The loss of mass in (6) slightly exceeded this value (65.6%), suggesting the total removal of the hemicellulose/lignin components, and with partial removal of other components. This interpretation agrees with the appearance of the material obtained at the end of processing—a cellulose-like fiber with a slightly yellow color. The results in Table 1 indicate the success of delignification with the use of 17.5% (w/v) aqueous NaOH for approximately 10–12 h treatment with wheat straw. This process is efficient, but heating for 10–12 h is an expensive process in terms of energy consumption. This may not be a problem for the large-scale process, but it can make small biomass processing units uncompetitive. Alternative conditions must be found for the small industrial units.

2.3.2 *Biomass delignification in alkaline conditions assisted by microwave heating*

Heating by microwave radiation is gaining strength due to the greater energy efficiency in heating, coupled with the catalytic action in carrying out some reactions promoted by the activation of excited vibrational and rotational states at the molecular level. In many cases, this gives rise to reaction rates, for example in hydrolysis and esterification, which are much higher than those obtained under thermal heating conditions. The ether bonds of lignin, having dipole moment (μ^-), should be especially susceptible to activation under microwave irradiation, making the delignification process much more efficient (Fan et al. 2017). In order to obtain microwave irradiation at practically continuous levels, temperatures higher than 100°C must be predicted, which requires vessels capable of withstanding pressures of at least 40 atm, and transparent to microwave radiation. At the laboratory scale, the use of Teflon wall vases is the solution, medium scale reactors can be used metal vases, capable of reflecting microwave radiation on their walls.

Microwave radiation is known to be particularly well absorbed by metal ions such as Fe^{3+} (Tsubaki et al. 2016), and its presence in the reaction medium may promote a secondary catalysis mechanism, improving the process. Some examples of accelerating the rate of hydrolysis of starch polysaccharides with the use of microwave radiation with the assistance of metalic salts are described (Li et al. 2001). Free radicals are characteristic of enzymatic degradation of lignin by peroxidases, enzymes that generate radical species. If iron is present, Fenton-type reactions are expected to occur (Pereira et al. 2012), they may assist in the hydrolysis of lignin.

Tests for the use of microwave radiation on a laboratory scale in the removal of base-catalyzed lignin/hemicellulose (NaOH), using wheat straw as a substrate, are presented in Table 3. As reference values for performing delignification, a sodium hydroxide concentration value in the range of 1–3% (w/v), about 20 times less than under thermal heating conditions, was chosen to promote delignification.

The results show a great superiority in relation to delignification with thermal heating, with efficiency equal to or higher than the calculated maximum theoretical value, with the use of only 2–3% NaOH (w/v) in periods of one hour. The properties of the final product are similar or superior to those obtained by thermal heating processing, with a fraction of the energy expenditure in obtaining it. It was found that experiments with sodium hydroxide in a concentration of 2 or 3% provide better final properties to cellulose (experiment 7, 8). The use of a small concentration of base also facilitates the removal of NaOH, requiring less expense with solvent in the washes. The use of metals is found to catalyze delignification (experiments 2, 5 and 9), however, in some cases, the residues of iron salts are adsorbed by the biomass to give it color, which may be undesirable. Its use depends on the final destination of the product.

Table 3. Delignification of wheat straw biomass by microwave heating with sodium hydroxide under different experimental conditions.

Experiment	[NaOH] % (w/v)	Time (min)	Temperature Program[1]	Mass removal (%)	Notes
Theoretical	–	–	–	41–63.0%	
1	1	60	PG 2	49.4 ± 1.7	Presence of Cu3+
2	1.5	60	PG 2	59.6± 0.8	Presence of Fe2+ (2x)
3	2	50	PG 1	44.7 ± 0.9	
4	2	60	PG 2	42.4 ± 0.7	
5	2	60	PG 2	45.3 ± 1.3	Presence of Fe2+ (1x)
6	2	60	PG 2	46.5 ± 0.9	Presence of Cu2+
7	3	50	PG 1	45.2 ± 0.5	
8	3	60	PG 2	55.9 ± 0.7	
9	3	50	PG 1	51.2 ± 0.9	Presence of Fe2+ (1x)
10	3	60	PG 2	60.8 ± 1.8	Rice straw
11	–	60	PG2	44.5	Rice straw, 3% KOH

Notes. 1. Program 1 (PG 1)—preheating period of 10 min at 140°C, followed by 40 min of heating at 160°C. Program 2 (PG 2)—10 min preheating period at 140°C, followed by 40 min of heating at 170°C. (1 x) ≈ 1 x 10^{-4} mol metal/50 mL.

2.3.3 Removal of amorphous cellulose

To prepare the MCC, a subsequent step is necessary to remove the amorphous cellulose fraction from the de-lignified cellulose. The classic conditions for preparing MCC from paper pulp are also suitable in the present case. Table 3 shows some experimental results of amorphous cellulose removal, when the raw material was wheat and rice straw cellulose, obtained by delignification using microwave irradiation.

The use of boiling 2M HCl for 15 min with delignified biomass are suitable reaction conditions for the hydrolysis of amorphous zones of high crystallinity cellulose obtained from grasses. The results shown in Table 4 involve the use of different conditions with different reaction times. For both rice straw and wheat straw, the weight loss of the biomass must reflect the complete removal of the amorphous areas of the cellulose together with the removal of parts of lignin that might not have been removed in the previous step. The boiling of 2M HCl, are conditions indicated to obtain the MCC from the biomass of cereal residues, a process similar to that used in MCC preparation from cellulose paper raw material. The color of the final product tends to be slightly yellow, in most cases due to traces of oligomers derivatives of lignin. In cases where the MCC has a high whiteness requirement, a bleaching process must be provided for. The use of whiteners such as H_2O_2 or peracids is particularly effective and can be done at several points in the processing described here. In the optimization of some

Table 4. Hydrolysis of the fraction of amorphous cellulose from cereal straw residues delignified to obtain MCC, by treatment with hydrochloric acid.

Experimental conditions					
Sample	Straw type	Time (min)	Temp. °C	[HCl] M	Weight loss (%)
1	Rice	15	100	2	28.6
2	Rice	20	100	2	35.7
3	Rice	25	100	2	37.4
4	Rice	30	100	2	37.5
5	Wheat	15	100	1	22.4
6	Wheat	15	100	2	30.0
7	Wheat	20	100	2	32.3

properties of the obtained MCC, final mechanical polishing treatments can be carried out, such as grinding in a roller or ball mill. This finishing process leads to a reduction and homogenization of the particle size. The final product can also be subjected to additional final grinds and/or coupled to particle size separation in an ultrasonic vibrating sieve tower, which is effective in adjusting the final particle size to produce the desired grades of MCC. In this way the product obtained will be as described in the list of characteristics for the European Pharmacopoeia MCC (European Pharmacopoeia 1999).

3. Nanocellulose crystals (CNC) and nanocellulose fiber (CFN)

More recently, a major research effort is being made in an attempt to obtain nanocellulose crystals and nanocellulose fibers from MCC from biomass or through techniques such as cell culture (Börjesson and Westman 2015, Löbmanna and Svagan 2017). The fact that these materials have a dimension on the nano scale, new properties are given to them enabling the construction of a set of new materials that are promising in different areas of science. For example, taking into account the mechanical properties of nanocellulose crystals, especially strong composites can be obtained that are possibly biocompatible. Nanomaterials also have a very high specific surface area, leading to the possible preparation of different colloidal dispersions. These have potential uses in fields such as the preparation of food, excipients and polymers for the controlled release of drugs, supercapacitors, and batteries, etc. Table 5 shows an approximate calculation of the surface area of the cubic particle versus its diameter in size, per gram of substance.

Of particular interest are colloidal dispersions that are usually formed with any material with a particle size between 10 to 100 Å (1 to 10 nm). The synthesis of nanocellulose crystals from filter paper or paper pulp is possible in a single step, with only water or slightly acidic water in a microwave oven at temperatures of about 230°C, with pre-set times heating for 10 min at 150°C, for total periods of time around 1 h. On cooling, these conditions

Table 5. Values for the edge size and total surface area of 1×10^{-6} m³ of a crystal shattered by mechanical means in a large number of small derived cubes.

Number of crystals	Edge dimension (m)	Total superficial area (m²)	Notes
1	0.01	0.0006	
8	0.005	0.0012	
64	0.0025	0.0024	
512	0.00125	0.0048	
//	//	//	
1.44E+17	1.91E-08	314.6	
1.15E+18	9.54E-09	629.1	
9.22E+18	4.77E-09	1258.3	Colloidal dimension
7.38E+19	2.38E-09	2516.6	
5.90E+20	1.19E-09	5033.2	
4.72E+21	5.96E-10	10066.3	
3.78E+22	2.98E-10	20132.7	

result in obtaining a transparent gelatinous material, in which the cellulose crystals have reached the colloidal dimension. These conditions are difficult to repeat and depend on the type of cellulose used, probably related to the different degree of crystallinity, the presence of ions and the microwave radiation pattern in the ovens or other unknown reasons.

More recently, the CNC is prepared from MCC by exfoliating the crystals through treatment with sulfuric acid in precise concentration and time. The basic process uses sulfuric acid in a defined set during a precise time of action. These acid hydrolysis conditions wear out outer layers of crystalline cellulose (MCC) decreasing in size until they reach nanometric dimensions. In the process, a considerable amount of sugar is released, along with traces of furfural. These can be collected and then fermented in ethanol. CNC particles can be prepared by treating MCC in 64% sulfuric acid solution (w/w) at 45°C for 45–60 min with constant agitation, followed by a rapid dilution of the acid by 1/10 with deionized water to stop exfoliation. The CNC can be isolated by centrifugation or dialysis. Nanocrystals tend to aggregate, and depending on the pH value, the use of ultrasound can be used to disperse them. In this way, the MCC from biomass residues may be a source of high quality crystalline materials and has a new potential for use—being the useful raw material for the preparation of CNC, a material that is being studied and will certainly originate new uses in several fields (Shojaeiarani et al. 2019, Kontturi et al. 2018). However, its detailed discussion is outside the scope of the present work.

CNF is more difficult to obtain from crystalline cellulose, as the fibers must be released from the crystal in an almost non-hydrolyzed form. Evidence that the process can be carried out with concentrated acid at low temperature occurs in decrystallization tests followed by immediate precipitation with alcohols (see Chapter 6). Tests using this method on crystalline cellulose samples by treatment with sulfuric acid at 0°C followed by precipitation with methanol gives rise to a white material, a process that is repeatable, generating a product similar to that observed in SEM images published in the literature for CNF (Zheng et al. 2014). A study by computer-assisted automatic tests is underway, with future publication in detail.

4. Antioxidant properties of lignin hydrolysates

In the delignification process, lignin is depolymerized by reactions catalyzed by a strong base, cutting the phenyl-alkyl ether bonds (Arai et al. 2019). Taking into account the structure of lignin, hydrolysis products must be phenolic substances, with a probable predominance of polyphenols, with some degree of sulfonation in the case of the use of sulfite in the process.

Polyphenols are substances disseminated among plants and to which strong antioxidant properties are attributed (Lingua et al. 2016, Dzah et al. 2020). Lignin itself is considered a substance with high antioxidant properties (Dizhbite et al. 2004, Mahmood et al. 2018). The concept of antioxidant is the subject of many interpretations and is evaluated using different analytical methods. Oxidation is the loss of electrons; whereas in the case of organic molecules, atoms of C, O, H or N dominate that have limited oxidation states, oxidation sometimes generates free radical species. In oxidation/reduction of metals such as manganese, for example, the metal can assume relatively stable ions from $Mn2+$ to $Mn6+$. Organic free radicals are very unstable, having a very high reactivity. Oxidation/reduction plays vital steps in metabolism and biological systems have developed strategies to confine the reactivity of these species. When released or formed by other means, such as electromagnetic radiation, species with a high energy content can originate, usually via interaction with the oxygen usually present, giving rise to species known as reactive oxygen species (ROS), examples of which are hydroperoxides, superoxide ($O2-$), and different radicals like $\cdot OH$ and $\cdot OOH$.

In different fields, ROS give rise to problems for the stability of both industrial products and fabrics made by industry. In products of daily use, foods are degraded, for example fats, juices and wines, products of daily use, perfumes and creams, and in the plastics and rubber industry, among many others. To mitigate the degradation of goods, substances that inhibit or eliminate these species—antioxidants—are added to quickly destroy any free radicals present. In the case of juices and wines the substance generally added is sulfite, which reacts quickly with oxygen, in a process in which it is expended, resulting in sulfate. When the sulfite ends, the antioxidant effects disappear.

The seeds are, in a simplified description, a reserve of nutrients, which include protein, lipid and sugars in the form of polymers, along with the DNA that conveys information for the birth of a new plant, when environmental conditions permit. There are known examples of plant seeds that have remained viable for periods of time between 2,000 to 30,000 years (Yashina et al. 2012). Lipids are particularly reactive with ROS species, such as atmospheric oxygen. This means that nature uses another technique, capable of efficiently preserving lipids in relation to the aggressiveness of oxygen that may eventually be transported through the seed membranes.

Photosynthesis occurs in the leaves of plants and involves numerous redox reactions. In these, the presence of species that destroy radicals, or antioxidants, is necessary to confine free radicals capable to complex enzymatic systems, otherwise the mechanisms existing there could be easily damaged.

Free radicals, if not stabilized, are very reactive species and prone to make the reaction of abstraction of H atoms or the formation of covalent bonds with close species, until they are either stabilized or annihilated. Two different mechanisms of antioxidant activity have been described above: the stoichiometric antioxidant, where a suitable substance reacts with the free radical destroying it, but at the same time, the substance is consumed. Examples are vitamin C or sulfite, used in the preservation of wines or juices against the residual presence of oxygen. Figure 1(a) shows the reaction of the sulfite with a radical species, capable of converting it to sulfate.

For long-term protection, these antioxidant species would not be able to protect, for example, unsaturated lipids in a seed against residual oxygen and, in this case, catalytic antioxidants are more suitable. These substances are easily found in fruits, flowers and seeds and sometimes have color.

Its mechanism of action is described in Figure 1(b), exemplified with 2,4-isopropyl, 5-t-butyl-phenol (1) substance that mimics the structural details that are found in natural antioxidants such as epicatechin shown in Figure 1(c). The substance (1) shares characteristics that allow it to work as antioxidant species: it has at least one or several aromatic rings, which by definition have a system of electrons π conjugated and spread across a carbon skeleton; a high number of hydroxyl groups (–OH). The aromatic part of the substance is protected by bulky organic groups that prevent access to the free radical. These characteristics are seen in all polyphenols, a fact that gives the same antioxidant properties.

The mechanism of action involves the arrival of a free radical, alkyl, aryl or ROS within the reach of the antioxidant substance. If R· abstracts one H· from an –OH group of the antioxidant, result in the annihilation of the radical resulting in RH and the formation of a antioxidant molecule free radical (step 1, Figure 1b). In this way, the –OH groups function as "antennas" capable of approaching free radicals. The antioxidant substance stabilizes the free radical by means of two characteristics that it presents—the first is the resonance of electrons π spread over an extended aromatic structure, usually of two or three carbon rings. To simplify (1), it has only one ring, but it also has alternative resources to stabilize the free radical, in this case obstructing the access of molecules to the confined free radical in a limited aromatic system.

a)

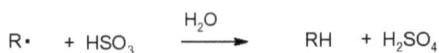

b)

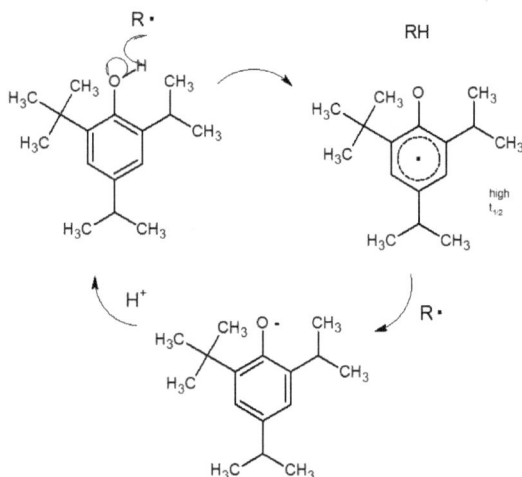

c)

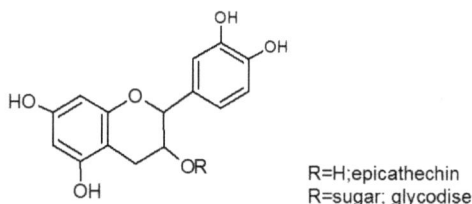

R=H;epicathechin
R=sugar; glycodise

Figure 1. (a) Stoichiometric antioxidant activity of sulfite in the destruction of oxygen; (b) catalytic antioxidant cycle presented by polyphenolic substances; (c) an example of a natural antioxidant, epicatechin present in tea leaves, a tissue where photosynthesis occurs.

This is achieved by placing bulky groups around the aromatic conjugation or other "antenna", to prevent physical contact with the radical. In nature, the use of sugars or methoxide groups around conjugated systems serves the same purpose. Aromatic conjugation is an important feature that stabilizes the radical, giving it time (usually measured as the half-life, t 1/2) to allow the approach of a second radical. This time delay is fundamental because a second radical species will again approach the O atom of the –OH group, converting the free radical into an anion, which by protonation regenerates the antioxidant substance in its original form (see Figure 1b). This mechanism is reminiscent of the annihilation of two radicals to form a covalent bond.

Lignin, being essentially a polyphenolic resin, by hydrolysis will give rise to a complex polyphenol mixture of low molar mass (Boeriu et al. 2004, Bailey et al. 2012). Thermal hydrolysis requires a concentration of NaOH up to 17%, versions using microwave irradiation require a lower base concentration in the range of 2–3%. The polyphenols in this strong basic medium are partially ionized and in neutralization with sulfuric acid to values close to pH 7 it produces two fractions—one soluble in water and the other that forms a fine suspension, which upon sedimentation produces a brown precipitate. The latter must consist of polyphenols with a ratio of carbon atoms to groups –OH less than approximately 1/5, therefore with more hydrophobic characteristics. The fraction that remains in solution are more polar phenolic substances or YAM. Both fractions of polyphenols derived from lignin have antioxidant properties, which means that they can be used in different applications. Human activity uses large amounts of synthetic antioxidants in different fields, in some cases toxic effects have been attributed to these artificial substances, while others in use have not yet been fully studied. From the human food chain in the stabilization of oils and fats, wines, even cosmetics, perfumery, pharmaceuticals, plastics and the polymer industry, its use is widespread. Derivatives of biomass can contribute to a greener society and, at the same time, ensure the supply of potentially safer antioxidants to different branches of industry.

The quantification of the antioxidant activity of soluble polyphenolic substances obtained from the degradation of wheat straw lignin, under conditions that allow obtaining MCC, was studied in detail and some results are provided in Table 4. The test selected to verify its antioxidant activity was the classic oxidation of 2,2′-azino-bis (3-ethylbenzothiazoline-6-sulfonic) or ABTS, to the colored cation ABTS+·, using as oxygen donor the H_2O_2, catalyzed by horseradish peroxidase (Putter and Becker 1985). Some versions use the colored radical ABTS+· preformed with persulfate oxidation, but there is some instability in the radical solutions (Boligon et al. 2014). The similarity of the results in both test trials and their conversion to antioxidant capacity equivalent to Trolox (TEAC) is used in the present evaluation (Re et al. 1999, Lingua et al. 2016). The activity of the diluted hydrolysates was compared with diluted red wine solutions and standard Trolox solutions using a spectrophotometric method, with ABTS+· radical ion monitoring at 660 nm.

The results of the antioxidant behavior of different antioxidants, with stoichiometric or catalytic properties are shown in Figure 2. The top of the Figure shows the behavior of two solutions, one of vitamin C in the ABTS test; using a concentrated solution of the same vitamin the complete inhibition of formation of the ABTS+· radical is obtained for 60 s (1), however by diluting this same solution in 1/100, a first phase of total inhibition occurs, followed, after 22s, by the formation of ABTS+· radical, observed by the development of absorbance at 660 nm (a-2). This example signals the presence of a stoichiometric antioxidant type (Rua et al. 2003).

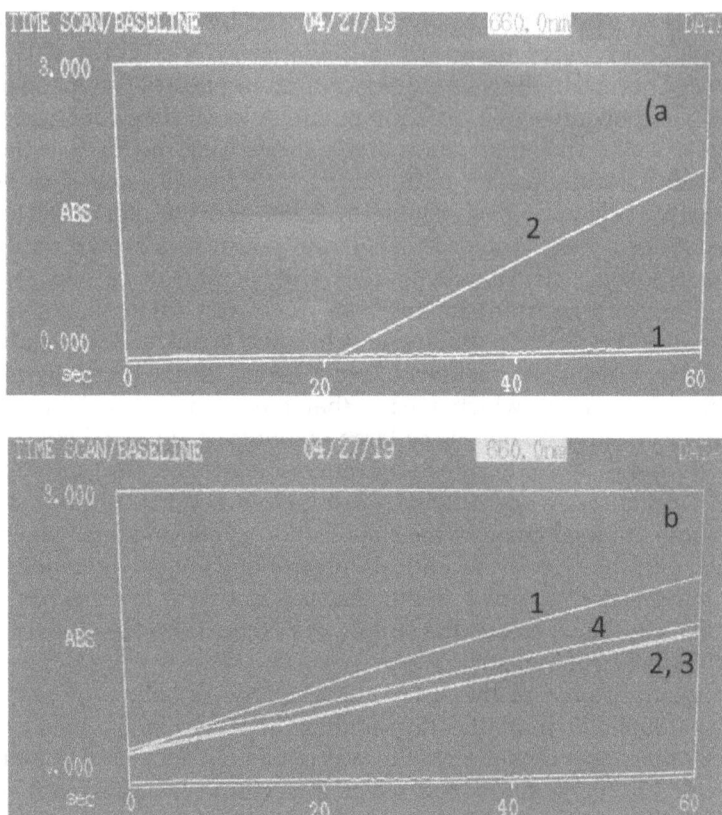

Figure 2. (a) Formation of the radical ABTS+· cation in the presence of Vitamin C solutions (1) observing total inhibition after 60 s; (2) with Vit. C solution diluted to 1/100, showing the inhibition of the formation of the cationic radical until 22s, followed by the beginning of formation of radical cation; (b) the same test with a blank (1), with a diluted Trolox solution (2 and 3) and with a sample of with KOH wheat straw hydrolysate, neutralized, filtered and diluted (4).

In the lower part of Figure 2, different tests of formation of cationic radical ABTS+· are shown: (a) with water, an example of a blank assay (1); with a standard Trolox solution, (2 and 3), and with a diluted neutralized soluble fraction of a wheat straw lignin hydrolyzate (4). As can be seen, the last solution has antioxidant properties. The quantification of the antioxidant activity of some of these extracts is shown in Table 6, using rice or wheat straw as starting material.

The results in Table 6 indicate that the hydrolyzates of soluble lignin from rice and wheat straw have antioxidant activity. The values found are lower than those found for a sample of red wine mixture, but close, in the range of 50–80%. Water insoluble precipitate test also shows antioxidant activity, but there is a difficulty in the evaluation due to the interference of solvents in the ABTS test.

Table 6. Experimental results of formation of the ABTS+· cation radical in the presence of neutralized, filtered and diluted cereal straws hydrolysates, using thermal and microwave radiation heating.

Sample		[Lignin] g/L	ABTS+· formation inhibition %	TEAC mmol/L	Notes
	Conditions				
1	Distilled water	-----	0.0	0.0	Blank
2	Diluted red wine[1]	----	89.9	160.0	Reference
3	Wheat straw 1 lignin; 17.5% NaOH	2.20	49.3	87.7	Thermal heating
4	Wheat straw 2 lignin; 22.5% NaOH	2.50	87.5	155.7	Thermal heating
5	Rice straw 1 lignin; 3% NaOH	0.75	67.3	120.0	Microwave heating (MW)
6	Rice straw 2 lignin; 3% NaOH	0.70	63.4	113.0	MW heating
7	Rice straw 2 lignin; 3% NaOH	0.83	72.0	128.0	MW; presence of Fe2+ ions

1. Mix of different red wines, diluted by 1/10, from the Souzelas region, Coimbra, Portugal, from the 2019 harvest.

These results clearly indicate the potential for using lignin hydrolysates or their sulfonate derivatives as natural antioxidants in different types of uses. For years, the final destination of lignin or its sulfonates has been elimination and more recently burning for energy. The pulp industry is a major producer of this type of materials. Thus, it is possible to make better use of those residues or others from straw treatment, in substances with strong antioxidant activity, being able to be used in a wide range of activities replacing synthetic antioxidants. Both soluble and insoluble substances can have a useful destination, the latter for example in the rubber, cosmetics, food industries, among others. The pharmaceutical activity and the health effects of these substances can also be valuable and a target for study.

5. References

Anwarab, Z., Gulfraz, M. and M. Irshad. 2014. Agro-industrial lignocellulosic biomass a key to unlock the future bio-energy: A brief review. Journal of Radiation Research and Applied Sciences 7: 163–173.

Arai, T., Biely, P., Uhliarikova, I., Sato, N., Makishima, S., Mizuno, M., Nozaki, K., Kaneko, S. and Y. Amano. 2019. Structural characterization of hemicellulose released from corn cob in continuous flow type hydrothermal reactor. Journal of Bioscience and Bioengineering 127: 222–230.

Bailey, R. W. and M. J. Ulyatt. 2012. Pasture quality and ruminant nutrition II. Carbohydrate and lignin composition of detergent-extracted residues from pasture grasses and legumes. New Zealand Journal of Agricultural Research 13: 591–604.

Boeriu, C. G., Bravo, D., Gosselink, R. J. and J. E. Van Dam. 2004. Characterisation of structure-dependent functional properties of lignin with infrared spectroscopy. Industrial Crops and Products 20: 205–218.

Boligon, A. A., Machado, M. M. and M. L. Athayde. 2014. Technical evaluation of antioxidant activity. Med. Chem. 4: 517–522.

Börjesson, M. and G. Westman. 2015. Crystalline Nanocellulose—Preparation, Modification, and Properties. Cellulose—Fundamental Aspects and Current Trends. Cellulose—Fundamental Aspects and Current Trends Edited by Matheus Poletto and Heitor Luiz Ornaghi Junior, Intech, 159–191.

Dizhbite, T., Telysheva, G., Jurkjane, V. and U. Viesturs. 2004. Characterization of the radical scavenging activity of lignins—natural antioxidants. Bioresource Technology 95: 309–317.

Dzah, C. S., Duan, Y. Q., Zhang, H. H., Wen, C. T., Zhang, J. X., Chen, G. Y. and H. L. Ma. 2020. The effects of ultrasound assisted extraction on yield, antioxidant, anticancer and antimicrobial activity of polyphenol extracts: A review. Food Bioscience 35: 100547.

Eichhorn, S. J. and R. Young. 2001. The Young's modulus of a microcrystalline cellulose. Cellulose 8: 197–207.

European Pharmacopeia. 1999. European Department for the Quality of Medicines within the Council of Europe, Strasbourg.

Fan, G. Z., Wang, Y. X., Song, G. S., Yan, J. T. and J. F. Li. 2017. Preparation of microcrystalline cellulose from rice straw under microwave irradiation. Journal of Applied Polymer Science 134: 44901.

Kontturi, E., Laaksonen, P., Linder, M. B., Nonappa, Groschel, A. H., Rojas, O. J. and O. Ikkala. 2018. Advanced materials through assembly of nanocelluloses. Advanced Materials 30: 1703779.

Kuznetsov, B. N., Sudakova, I. G., Yatsenkova, O. V., Garyntseva, N. V., Rataboul, F. and L. Djakovitch. 2018. Optimizing single-stage processes of microcrystalline cellulose production via the peroxide delignification of wood in the presence of a titania catalyst. Catalysis in Industry 10: 360–367.

Laka, M. and S. Chernyavskaya. 2007. Obtaining of microcrystalline cellulose from softwood and hardwood pulp. Bioresources 2: 583–589.

Li, K. L., Xia, L. X., Li, J., Pang, J., Cao, G. Y. and Z. W. Xi. 2001. Salt-assisted acid hydrolysis of starch to D-glucose under microwave irradiation. Carbohydrate Research 331: 9–12.

Li, M., He, B., Li, J., Li, J. R. and L. H. Zhao. 2019. Physico-chemical characterization and comparison of microcrystalline cellulose from several lignocellulosic sources. Bioresources 14: 7886–7900.

Lingua, M. S., Fabani, M. P., Wunderlin, D. A. and D. A. M. V. Baroni. 2016. From grape to wine: Changes in phenolic composition and its influence on antioxidant activity. Food Chem. 208: 228–238.

Löbmanna, K. and A. J. Svagan. 2017. Cellulose nanofibers as excipient for the delivery of poorly soluble drugs. International Journal of Pharmaceutics 533: 285–297.

Mahmood, Z., Yameen, M., Jahangeer, M., Riaz, M., Ghaffar, A. and I. Javid. 2018. Lignin as natural antioxidant capacity. trends and applications. In Lignin—Trends and Applications. Intech Open.

Moon, R. J., Beck, S. and A. Rudie. 2013. Cellulose nanocrystals—A material with unique properties and many potential applications. *In*: Postek, M. T., Moon, R. J., Rudie, A. and M.A. Bilodeau (eds.). Production and Applications of Cellulose Nanomaterials. Tappi Press, GA, USA.

Ohwoavworhua, F. O. and T. A. Adelakun. 2010. Non-wood fibre production of microcrystalline cellulose from sorghum caudatum: Characterisation and tableting properties. Indian Journal of Pharmaceutical Sciences 72: 295–301.

Pereira, M. C., Oliveira, L. C. and E. Murad. 2012. Iron oxide catalysts: Fenton and Fenton-like reactions—a review. Clay Minerals 47: 285–302.

Putter, J. and R. Becker. 1985. In Methods of Enzymatic Analysis. Vol. 3. Eds H. U. Bergmeyer, VCH Publications, Weinheim, Germany.

Ramezani, M. G. and B. Golchinfar. 2019. Mechanical properties of cellulose nanocrystal (CNC) bundles: Coarse-grained molecular dynamic simulation. J. Compos. Sci. 3: 57.

Re, R., Pellegrini, N., Proteggente, A., Pannala, A., Yang, M. and C. Rice-Evans. 1999. Antioxidant activity applying an improved ABTS radical cation decolorization assay. Free Radical Biology and Medicine 26: 1231–1237.

Rua, M. S., Ribeiro, T. C., Félix, M. F., Costa, M. C. and J. M. T. B. Varejão. 2003. Contribution to the study of the antioxidant activity of Portuguese wines. The Proceedings of 6th Meeting of Portuguese Food Chemistry, Lisbon, Portugal.

Shojaeiarani, J., Bajwa, D. and A. Shirzadifar. 2019. A review on cellulose nanocrystals as promising biocompounds for the synthesis of nanocomposite hydrogels. Carbohydrate Polymers 216: 247–259.

Tsubaki, S., Oono, K., Onda, A., Yanagisawa, K., Mitani, T. and J. Azuma. 2016. Effects of ionic conduction on hydrothermal hydrolysis of corn starch and crystalline cellulose induced by microwave irradiation. Carbohydrate Polymers 137: 594–599.

Wu, X., Moon, R. J. and A. Martini. 2014. Tensile strength of Iβ crystalline cellulose predicted by molecular dynamics simulation. Cellulose 21: 2233–2245.

Yashina, S., Gubin, S., Maksimovich, S., Yashina, A., Gakhova, E. and D. Gilichinsky. 2012. Regeneration of whole fertile plants from 30,000-y-old fruit tissue buried in Siberian permafrost. Proceedings of the National Academy of Sciences of the United States of America 109: 4008–4013.

Yavorov, N., Valchev, I., Radeva, G. and D. Todorova. 2020. Kinetic investigation of dilute acid hydrolysis of hardwood pulp for microcrystalline cellulose production. Carbohydrate Research 488: 107910.

Yue, X., He, J., Xu, Y., Yang, M. X. and Y. J. Xu. 2019. A novel method for preparing microcrystalline cellulose from bleached chemical pulp using transition metal ions enhanced high temperature liquid water process. Carbohydrate Polymers 208: 115–123.

Zheng, Q. F., Cai, Z. Y. and S. Q. Gong. 2014. Green synthesis of polyvinyl alcohol (PVA)-cellulose nanofibril (CNF) hybrid aerogels and their use as superabsorbents. Journal of Materials Chemistry A 2: 3110–3118.

Chapter 5

Carbon Fiber Analogues by Fusion of Biomass Polymers

1. Introduction

Biomass has in its elemental composition about 50% carbon. Life, as we know it, is made possible by the multiple ways in which carbon chemically bonds with itself and with most elements of the periodic table, making it possible to build complex structures. When any type of biomass is heated to high temperatures, the same material is obtained—coal, which is composed mainly of carbon. There are many different forms of carbon, each having very different properties (Baseri et al. 2012). Carbon-related materials have been and are historically relevant to human civilization, their use remains important in various activities. The classic uses can be consigned as a source of energy for food preparation, as a source of energy for metallurgy, and an agent with a reducing capacity, among many others. Some countries, like South Africa, with rich deposits of fossil carbon in the subsoil, convert them into a mixture of carbon monoxide and hydrogen, the synthesis gas or syngas, to distribute or to be used as raw material to generate all types of hydrocarbons that modern life offers the consumer. The disadvantage is a net contribution of carbon dioxide emissions to the Earth's atmosphere, which are beginning to destroy the global ecosystem.

More recently, new carbon alloforms have been developed, examples are fullerenes, carbon nanotubes, graphene, among others. These new variants promise a range of new materials in the near future, examples are supercapacitors, resistant electrodes, lighter and stronger carbon fibers and new electrical conductors, the possible development of superconductors at room temperature, new textiles with different properties and the possibility of manufacturing composite materials with different properties. Until recently, new carbon materials were made exclusively through synthetic routes, however, more recently there is an obvious interest in trying to incorporate or start making these materials from biomass residues. In recent years, there has been an increase in the research effort for using biomass as a suitable raw material for the preparation of new carbon materials.

In any high temperature biomass treatment process, such as pyrolysis, three products are obtained: charcoal, a gas mixture that under condensation conditions gives rise to two fractions—a viscous oil—bio-oil, and a gas fraction, the synthesis gas. Experimental conditions exist to optimize any of the fractions generated in the process.

2. Charcoals

Charcoal can be prepared from almost all types of biomass by pyrolysis or extracted from the subsoil as a mineral. Both originate from biomass, but one has undergone a long stay underground under conditions such as high pressure and/or temperature, leading to major transformations in its composition, essentially lowering the elementary ratios of O/C and H/C, relative to the starting material. The coal mineral is generally divided into sedimentary or magmatic types, with properties that differ considerably, an example of the latter group being anthracite.

Charcoal is characterized by the occurrence of major transformations in the structure of biomass—they usually have varied structures with different ones reflecting the origin of the original biomass component and, therefore, can be classified as heterogeneous carbon materials. Wood charcoal is made from different plant sources by pyrolysis, where it undergoes a weight loss of 25 to 70%, depending on the preparation conditions; namely temperature, oxygen content and treatment time. Typically, higher temperatures result in greater carbon loss, but resulting in a high purity end product (Wang et al. 2016). Oxygen must be avoided during the process, and it is desirable that the decomposition is done in an inert atmosphere. Decomposition temperatures start at around 380°C for biomass components with less stability such as hemicellulose; components like crystalline cellulose are carbonized at much higher temperatures, in the range of 500–600°C. In the global process, a temperature below the incandescent temperature ($\approx$ 700°C) is desired. For the preparation of activated carbon, heating to 800–900°C in a carbon dioxide or water oxidizing atmosphere, induces molecular rearrangements with high porosity formation in the coal.

Different preparation techniques can be visualized, on a laboratory scale, in a high temperature oven under an inert atmosphere, with the aid of thermal heating. In commercial plants the partial burning of biomass can provide the heat to reach the necessary temperature. In this case, part of the biomass is burned in a closed space with controlled oxygen concentration for the fraction of the stoichiometric combustion need. This allows maintenance at a high temperature while carbonization progresses. Coals, whether minerals or wood, are sold under different particle sizes, essentially as fine powder (PC's) or in larger particles, granular carbon (GC's).

The purity of charcoal can be accessed by the fixed carbon content, which represents approximately elemental carbon content in the sample. The remaining fractions are oxides of inorganic metals and minerals, a content that can be evaluated by quantifying the ash in calcination.

As a model for understanding the processes that take place in biomass at high temperature, it may be useful to think about the types of carbon bonds. Most new carbon materials, including carbon fiber, have good electrical conductivity. Electrical (e–) or hole (h+) conductivity requires the presence of conjugated double bonds (C = C) in a single carbon carbon (C-C) backbone. This also leads to the spectroscopic properties of the carbons and, specifically, their black color. Different domains are involved, which have long conjugation chains involving electrons π, with the probable presence of aromatic parts.

As the original carbon bonds in biomass are predominantly simple bonds, these assume tetrahedral geometry (sp3), in connection with itself or with different heteroatoms. Tetrahedral carbons prevent resonance and are not conductors of electrons or gaps.

Thus, carbonization can be described as a thermal process involving high levels of energy at the molecular level that generate breakage of covalent bonds, formation of radical species, which yield to rearranging the components of biomass, first by elimination reactions releasing oxygen atoms, nitrogen and hydrogen in the form of water and nitrogen oxides with the concomitant formation of a greater number of carbon double bonds, and in later stages, at higher temperatures by the release of a higher content of heteroatoms, inorganic species and metal oxides. In this process, aromatic carbon compounds are the most thermodynamically favorable products and are reorganized in phases related to asphaltenes. These are already considered to have an aromatic nucleus with at least 7 rings, up to much higher values and surrounded by the presence of alkyl groups (Ferreira 2017). The trend of π-stacking formation begins to emerge in these, giving rise to parallel layers of fused aromatic rings. By condensing these units at a higher temperature, the final domains that are generated may be folded graphene-like sheets, graphite, turbostratic graphite (Binder et al. 2017), which are generally imperfect and present as heterogeneous local domains. With higher temperature treatments, the presence of flat sheets of fused C6 rings can begin to form phases such as graphite perfectly stacked in a crystal structure or turbostratic graphite when a disorderly stacking occurs between graphene-type carbon sheets. From these materials under conditions of a very high temperature, in the range of 2,000 to 3,000°C, under an inert atmosphere, the final rearrangement may occur to the graphitization of carbon.

In the preparation of charcoal, the above process is interrupted at an intermediate stage, having dissimilar phases or domains not yet perfect that can originate a three-dimensional aromatic macromolecular structure rich in aromatic domains both with nanopores and with micromolecular fissures (Mochidzuki et al. 2003), forming a less dense material, with weak mechanical properties, that is, with low breaking stress and elastic modulus.

The porosity is particularly interesting because it gives the carbon materials a high surface area, associated with the very good carbon adsorption properties, making them suitable for the preparation of active charcoals. The pore size differs in micropores (< 2 nm), mesopores (3–40 nm) and macropores

(> 50 nm), depending on the thermal processing and the biomass source of material. These have a wide range of applications from water purification: in the removal of metals and/or organic molecules such as polycyclic aromatic hydrocarbons (PAHs), dyes in textile industry effluents, with the advantage of simple regeneration by heating at high temperature. Its use is widespread in many other fields such as pharmaceutical, dietetic, whiskey preparation, and the pharmaceutical industry, among many others.

2.1 Bio-oil

Bio-oil is the oil obtained in the degradation of any biomass through high temperature treatment, consisting of molecules that are released in the formation of coal. It is a very complex mixture of substances originating from biomass components that condense at room temperature from high temperature processing. Thus, components derived from hemicellulose such as acetic acid and furfural; lignin derivatives such as phenol and substituted methoxy phenols, substances originated from cellulose degradation, such as levoglucosan and 2-hydroxymethylfurfural, which are associated with a large number of other substances, with low molar masses, protein degradation products, among others. A portion of asphaltenes with higher molar mass may also be present, usually originating from lignin and incorporating any of the other components of the biomass.

Efforts have been made in recent decades to find useful end uses for bio-oil, since it is a form of liquid biomass, with high energy density, with all the advantages of liquid fuels, such as easy transportability, storage, etc. Its use as a biofuel is a possibility. Unfortunately, it has a very high O/C ratio in its composition, compared to normal fuels, which must be reduced by chemical means. The main oxygenated substances present are carboxylic acids, such as acetic acid from hemicellulose and oxidation by atmospheric oxygen of aldehydes obtained in the decomposition, making it corrosive. At the same time, it is easy to polymerize through condensation reactions between aldehydes and alcohols. A great effort is being made to stabilize it, through techniques such as reform, hydroformylation, disproportionation, catalytic hydrogenation and esterification.

Bio-oil production units are active in various parts of the world, the USA has the largest number, including portable units capable of converting the local accumulation of biomass and transporting it in the form of bio-oil (Mirkouei et al. 2016). Biomass residues are spread everywhere, an improvement in this process is desirable. As an example, Spain and Portugal could benefit from the conversion into bio-oil of the high quantity of olive tree cuts and the residual biomass obtained in the olive oil extraction process.

2.2 Syngas

The non-condensable gaseous part in biomass pyrolysis is mainly composed of a mixture of carbon monoxide (CO), hydrogen (H2) and traces of methane

(CH4). Since biomass pyrolysis is done under aerobic conditions, the significant nitrogen content is usually the main component. It is a gas mixture with relatively high energy content, mainly due to the presence of CO, and capable of being transformed into substances for the market such as liquid hydrocarbons such as gasoline, diesel or even waxy materials, through the Fisher-Tropsch process. South Africa is a good example of a country where fossil coal reserves are the raw material for preparing liquid fuels. Another useful product prepared from syngas is methanol, in a process in which it is necessary to make the stoichiometric adjustment of the CO/hydrogen content, being then converted to methanol:

$$2\ H_2 + CO \rightarrow CH_3OH$$

The reaction is carried out with the aid of different catalysts, the most common copper oxides at pressures as low as 2 to 40 atm. The Shift reaction, whose equation is shown below, can be used in the availability of heat and carbon to adjust the hydrogen/carbon monoxide ratio:

$$C + H_2O \rightarrow CO + H_2$$

This reaction requires high temperature, since carbon sources can be any type of plant biomass, but they can also include other types of biomass, such as animal and human manure. Nowadays, efforts are being made to extend the human presence beyond the Earth, these will require total biomass recycling capacity and this process can help, in a holistic approach, to obtain and recycle all resources. Converting syngas to methane may be another opportunity, as the process is well known.

3. From heterogeneous biomass to pure carbon holoforms

The last decades have seen the emergence of new forms of carbon materials, of homogeneous phase, some examples are carbon nanotubes, simple or in the form of multiple tubes, graphene, Buckyball's, etc (Hussein et al. 2020, Rubel et al. 2019). These are high purity substances related to graphene, other examples that have been known for a long time are graphite or turbostratic graphite, in which sheets of graphene are stacked by π-stacking. Some of these new forms of carbon have excellent electrical and mechanical properties making them very desirable for the preparation of new materials, which were quickly acquired by the industry.

More recently, studies on the use of residual biomass as a raw material for the production of new carbon materials, such as porous carbon fibers (PCFs), spheres, etc., have begun. Processing usually involves pretreatments with inorganic salts, acids or bases, or all of them, followed by thermal carbonization at high temperature. Both residual biomass and pulp, defibrillated cellulose or more crystalline cellulose can be used (Zheng et al. 2017). The process follows the sequence already formulated above, which corresponds to the sequence, biogas + bio-oil and higher fixed carbon materials, eventually resulting in graphite.

Crystalline graphite has long been known as metamorphic rock. It presents a layered structure of graphene-like sheets stacked by π-stacking. Graphite thus has a planar structure. Each sheet has an extended network of sp2 carbon with a thickness of a single carbon atom, conducting electricity, in the transverse direction of the sheet is a very strong material in terms of mechanical properties. Towards the sheets they can slide relative to each other, giving properties such as lubricity. A known way of obtaining graphite from coal is to heat it to temperatures in the range of 2,000–3,000°C under argon (nitrogen at this temperature reacts with carbon to produce nitriles) to volatilize remaining imperfections resulting in a final product of greater purity.

Carbon materials that have not yet reached this degree of purity have multiple phases–being called heterogeneous phase–and are easier to produce from biomass. The manufacture of this pure carbon from biomass is a challenge.

Carbonization of biomass at lower temperatures can generate layered domains, such as graphite, along with domains similar to unstacked graphene. The result is a heterogeneous structure in which a large number of defects prevent perfect stacking, being richer in turbostratic graphite. The graphite itself can be modified to a turbostratic form by simple extensive grinding or chemical exfoliation and reduction of the corresponding particle size (Gao et al. 2003). In this form, good electrical conductivity properties together with a high surface area are characteristic. Imperfections in turbostratic graphite or graphene-like carbons can originate in the residual presence of stabilized metals, inorganic ions or sp3 carbons. There may be a lack of carbon atoms to complete the structure, as it occurs in the degradation of sugars in C5 that tends to form cyclopentadiene rings, unable to give continuity to growing aromatic rings and adding an allyl position in structures similar to graphene. Electrical conductivity, allows for the presence of conjugated connections to begin to appear. Since the domains are physically separated, the general electrical conductivity requires good electrical contact between adjacent domains. The presence of allylic hydrogens in positions close to the extended conjugation systems makes carbon materials susceptible to oxidation by oxygen. In its presence, it is expected that the electrical resistance and color will be degraded over time.

4. Carbon fiber

In the most homogeneous carbon phases, most of the materials made in a synthetic way have excellent electrical and mechanical properties making them desirable for the preparation of new materials, which have been quickly acquired by the industry. Carbon fiber is an example of success, which is a material with very strong mechanical properties that can give rise to the preparation of very strong composites that can be used in a wide variety of fields (Liu and Kumar 2014).

The current process for making carbon fiber uses the acrylonitrile monomer as a starting point. Its polymerization in the liquid phase produces polyacrylonitrile (PAN) which is transformed into fiber by a spinning receptor technique (spininng) that forms the precursor (Yusofab and Ismailab 2017). This is subject to a sequence of the process, including cyclization, oxidation, a set of different heat treatments at increasing temperature, chemical treatments (Huang and Zhao 2016), sometimes with applied traction forces leading to the final process of "graphitization" at very high temperature. It produces a material with very high mechanical properties in terms of tensile strength and Young's modulus (Lau and Hui 2002). Structurally, carbon fibers are essentially turbostratic graphite, with a predominant sheet alignment in the direction of the fiber (Cao et al. 2018). The turbostratic graphite has very high tensile strength and Young's modulus value, since its graphene-like sheets are twisted and can even be positioned on axes perpendicular to each other, giving resistance to two dimensions. There are many types of carbon fiber and techniques for preparing it, and its field of application is spreading rapidly.

The price of carbon fiber is high, reflecting the complexity of the preparation and the price of the initial raw material, acrylonitrile, which accounts for about 50% of the cost. The precursor is expensive, which makes the preparation of these materials expensive and hence limits their widespread use.

A highly desirable goal is to try to use residual biomass in the manufacture of similar materials, such as carbon fiber, or to add biomass components to the existing manufacturing process to make it competitive. Figure 1 shows the sequences of chemical reactions taking place in the preparation of PAN carbon fibers and a comparison with a structure of the biomass components, mainly those rich in C6 rings, such as cellulose and lignin. PAN precursors are made by polymerization of acrylonitrile producing a fused polymer which on oxidation can be described as a fused ring of pyrido [2,3-d] pyridine (3, see Figure 1). These are triggered by oxygen oxidation (3), subsequently forming the corresponding N-oxides (4), which in the high temperature treatment start to release water and nitrogen oxides forming ring fusion structures (5). The process continues to grow at higher temperatures to give rise to graphene-like structures (see Figure 1). An analogy reaction path is elaborated in Figure 1 starting with cellulose.

Both in the PAN-CF carbon fiber precursor and in the synthetic route from biomass, a cyclic structure with the presence of a heteroatom forms the starting point; in biomass, approximately 80% of the fixed carbon forms structures in which a hexagonal ring is present. Thus, starting with cellulose, the acid/heat treatment promotes the elimination reactions of two water molecules per unit of glucose, producing units of 3-keto, α, β-unsaturated pyranose (7, see Figure 1). The reactivity of the latter is high, and at the temperatures used, keto-enolic and addition condensation reactions begin to produce the polymers with fused pyran rings. Pyran has characteristics

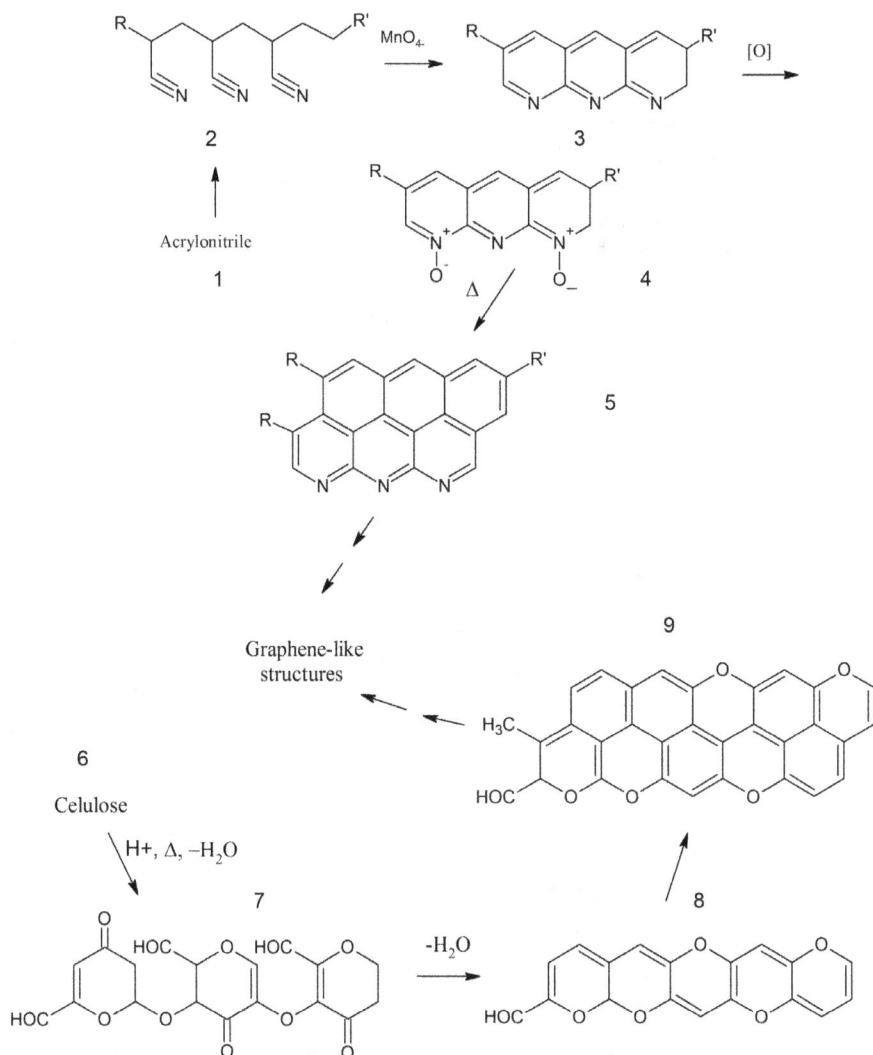

Figure 1. PAN synthesis reaction pathway (1–5) and a comparison with the possible aromatization pathway from cellulose fibrils (6–9) for the preparation of fused aromatic rings, possible precursors of graphene-like substances.

of aromatic compound, considering the hydridization of oxygen as sp2 and the easy formation of the cation by loss of H+. With heat treatment in more energetic conditions, the expulsion of oxygen facilitates the fusion of several cyclic pyran chains with water release to produce aromatic fused ring systems (9, see Figure 1). These substances increase the degree of aromaticity, since they are the product favored thermodynamically, starting to form substances related to asphaltenes. The continuation of this process increases the molar mass and extends the resonance by successive addition of rings, to produce substances similar to graphene. Some problems can arise when using biomass, such as the presence of hemicellulose, which with its structure of sugars in C5, can be decomposed into furans, these can be inserted or ejected into the structure of fused rings in formation. One of the routes is its incorporation into the rings in the form of cyclopentadiene, which makes planarization difficult due to the difficulty in introducing a sp3 carbon into the chain. The same carbon will form an allylic position, known for its relatively easy oxidation. Thus, hemicellulose may in some circumstances cause defects in the ring fusion process shown in Figure 1. Other groups, such as the propane alkyl part of lignin, must undergo rearrangement that allows their release, together with methoxide groups, as low molar weight substances, since a mechanistic pathway is not easily visualized for its incorporation into the aromatic superstructure.

While biomass as a whole is being studied for the preparation of carbon materials, the incorporation of lignin in the PAN-CF process has been tested. This is due to the large availability of lignin as a waste originating in the pulp industry, and largely burned. The preparation of carbon fiber (CF) with lignin incorporation is partially possible (Khalid et al. 2016, Xuefeng et al. 2018). To this end, the presence of aromatic rings contributes, however the presence of some degree of crosslinking, causes problems in their use in the PAN-CF process at levels greater than 20%, in the preparation of the precursor CF. Crosslinking tends to lower the oxidation capacity of the precursor, which is necessary for the progress of the process (Jiang et al. 2018). Additional chemical treatment techniques allow to partially overcome the problem and are being studied.

Direct lignin graphitization is possible by thermal heating at 850°C with catalysts such as nickel acetate, producing a high degree of crystallinity (Kubo et al. 2003). The mixture of cellulose and lignin has been successfully used as a raw material for the carbon fiber precursor by modified spinning techniques using ionic liquids such as [DBNH] OAc in the preparation of the precursor, under conditions other than the classic method, to produce similar precursor properties (Byrne et al. 2018).

The manufacture of carbon fiber analogs or other special carbon holomorphs from biomass residues is highly desirable, being more difficult than that of the components, in pure form described above. Further studies are needed to generate new techniques that are effective with the use of biomass.

As previously mentioned, carbon fiber made it possible to prepare materials with excellent mechanical properties that are very strong and low weight, essentially in the form of composites with, for example, epoxy resins. In these composites, the carbon fiber or the capillary nanotubes have the role of having a high tensile strength and Young module (see Table 3 at the end of this chapter), with the resin being the binder capable of joining and molding the fibers to the desired final shape. The mechanical properties of cellulose macrofibrils with a high content in the crystalline fraction are similar or even greater than carbon fibers (see Table 1, Chapter 4). These can be obtained from various sources of biomass with the advantage of being able to reach lengths of several meters, making them candidates in the preparation of composites with properties superior to those already available. An obstacle to the use of MCC or CNC in the preparation of composites is the highly hydrophilic character of their surface, rich in hydroxyl groups, making compatibility with more desirable synthetic resins, of the epoxy type, difficult to establish. This fact leads to the subsequent easy delamination of the formed composites. Surface acetylation is a possible technique to overcome this difficulty (Gan et al. 2017). The partial carbonization of cellulose fibers (for example CNF) will produce a more hydrophobic surface and compatibility with epoxy resins. This can give rise to the preparation of composites with good mechanical properties similar to those obtained by using carbon fiber and with less electrical conductivity, which is sometimes a problem. This will require new techniques to prepare substances derived from biomass with modified properties.

5. Hydrothermal carbonization (HTC process)

A new biomass carbonization technique has recently emerged and is beginning to gain visibility in the preparation of different forms of carbon, such as hydrochars, with potential for use in several fields (Luo et al. 2019). One of the new techniques is the biomass carbonization at high temperatures in water (HTC). The biomass is heated in the subcritical phase of the water, at a temperature in the range of 150–380°C, that is, at temperatures much lower than those used in classic carbonization, with residence times in the range of 2–25 h. The process has some desirable characteristics, rendering the drying and eventual grinding of biomass steps unnecessary the possibility of using a wider range of types of biomass, including manure and general waste, is less demanding in energy consumption, tends to generate less content of bio-oil and syngas. As a disadvantage, high temperature water requires a vessel capable of withstanding pressures of 12 to 50 atm, or even more, depending on the water temperature. Metal catalysts or the use of microwave radiation can make the process even more efficient and less energy consuming.

In the HTC process, it is common to identify two active mechanisms of action in parallel. One is the "soluble route", which refers to substances formed by heat treatment that diffuse in the aqueous medium. In this group, the dissolution of mineral salts and substances easily extractable from the decomposing biomass, such as hemicellulose fractions and sugars from their hydrolysis are expected. The other process that occurs in parallel are reactions in the solid phase of the biomass, generally known as the "solid path" mechanism (Zhang et al. 2018). This happens mainly in the cellulose and lignin phases.

The two processes, as occurs in classic thermal carbonizations, lead to the formation of three main products: coal, fraction of bio-oil poorly soluble in water and syngas. The charcoal formed in this process tends to have a lower fixed carbon value than those obtained by classical methods. The raw biomass has approximately 43% of the fixed carbon (dry mass) and in the charcoal obtained in the HTC process values up to 73.0% can be reached. The final value depends on the experimental conditions, the fact that carbonization is done in the presence of water may stabilize functional groups that have oxygen, making its removal more difficult. The charcoal obtained in the HTC process has high hydrophobicity and is a brittle solid, with a high porosity structure with large surface areas. The carbonization mechanism on HTC is the focus of intense debate, a proposal is formulated in the following section, supported by a large number of experimental results. An important note is that HTC occurs specifically in neutral or acidic aqueous solutions. Neutral aqueous media can become acidic by the action of hemicellulose, through the hydrolysis of its acetyl groups, which favors carbonization. The presence of small amounts of base leads to a different mechanism that favors hydrolysis and may prevent carbonization. If hydrolysis is favored, oligosaccharides, polyphenols and MCC and CNC become the main products, at least in the temperature range below 220°C.

HTC carbonization shows aspects that are not seen in the classic thermal process. The most significant is obtaining materials derived from biomass in the form of spheres, commonly referred to as the spherification process. The spheres observed have dimensions in the meso and macro scales, with the use of direct biomass, without any purification, they seem to form preferentially on the surface of cellulose macrofibrils, especially in places where it has suffered cuts. Carbonization studies of sugars under HTC conditions generate carbon products with high levels of spherification, mainly of nano diameter, examples are the use of sucrose at a temperature of 200°C (Romero et al. 2014).

The spherification of biomass requires the formation of an intermediate liquid with high surface tension, in an aqueous medium, a force that leads to the minimization of surface energy, gaining the liquid derived from biomass a spherical form. Thus, the liquid will have to have a high hydrophobicity, and the likely starting material that leads to the formation of such liquid is the degradation of hemicellulose and lignin, forming substances that apparently "melt", forming an intermediate liquid. The liquid subsequently polymerizes and undergoes reactions that lead to its transformation into a solid, a perfectly spherical form of carbon with high porosity, with values of surface area reaching values as high as 3100 m2/g (Romero et al. 2014). Aldehydes are reactive groups that in the presence of acid or base can give rise to a range of condensation reactions. Monosaccharides are aldehydes or ketones, in their pure form they are very soluble in water and hydrophilic, which points to the reductive elimination of hydroxyl groups and intermediate formation of a liquid mixture, which advances in subsequent condensation/elimination steps begins to generate materials hydrophobic solids.

The properties mentioned for the microspheres may allow its use as adsorbents, chromatography stationary phase materials, supercapacitors, etc. Figure 2 shows an example of spherification image on the surface of ground wheat straw macrofibrils, with spheres in the range of 5 to 50 μm, obtained by HTC at a temperature of 210°C, without any preliminary treatment.

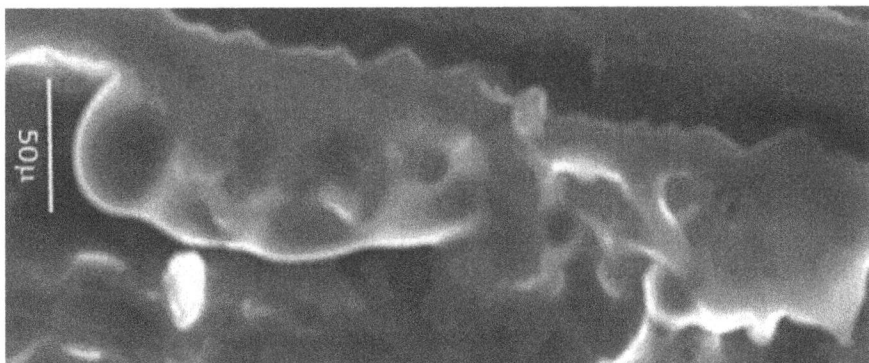

Figure 2. Example of formation of carbon spheres on the surface of cellulose macrofibrils in a wheat straw biomass sample submitted to the HTC process (see text for details).

6. Carbonization at very low temperature (VLTC)

The methods described above for preparing materials derived from carbonized biomass always involve the use of high temperature treatment. These are the rules among the techniques described in the literature today. The HTC process is a new means of processing biomass with the potential to add value to a wide range of biomass materials, suffering as its main disadvantage the need for vessels with reinforced walls to withstand high pressure and temperature conditions. However, the method shows that even with temperatures lower than those used in thermal carbonizations it is possible to obtain coals. The use of microwave radiation on HTC produces the same result, and lower temperatures can be used in the range of around 180–220°C, in the presence of diluted acid. In this process, more elaborate reactors requiring materials transparent to radiation, and capable of withstanding medium pressures make the process viable in small-scale laboratory experiments but make it difficult when a large-scale biomass is in process. The HTC process can be done in the laboratory using filter paper or biomass in a microwave oven with controlled temperature, depending on the model, capable of reaching a temperature of 240°C and resisting 40 atm. In the treatment of filter paper under conditions such as water treatment initially at neutral pH, and at a temperature of 180°C, the characteristic brown color of monosaccharide condensation products begins to be formed together with some degree of paper carbonization. Ideally, it would be much more advantageous if carbonization could be done at even lower temperatures, close to room temperature, and in reactors without the need for pressurization. When a small amount of acid is added to the HTC process the same result of carbonization of filter paper can be achieved at lower temperatures, close to 100°C. Similar effects have been observed over the years being reminiscent of old observations, such as the effect of concentrated sulfuric acid on sucrose at room temperature. This and other assays share the fact that they are rarely repeatable and difficult to control, and thus have aroused little interest over the years. However, if similar conditions could be controlled, it would allow the processing of high volumes of biomass, requiring little energy for carbonization. The understanding of the carbonization mechanism that occurs in the aqueous phase in the HTC process, can give indications on what parameters can be modified to make biomass carbonization at low temperature viable.

Non-thermal carbonization (T < 150°C) is not described in the literature, either for biomass or for any of its components. The use of acids, usually sulfuric acid, as substances capable of assisting the process is mentioned, followed by treatment at high temperatures, T > 200°C (Jawad et al. 2019). Detailed observation shows that the spherification mechanism that occurs in the HTC process in aqueous phase, sugars are converted into mixtures of hydrophobic substances of low molecular weight, and are capable of explaining the observed surface tension effects that lead to the formation of liquid spheres. This same liquid originates in more advanced stages of the reaction, porous carbon solids. The fixed carbon value of the starting materials, sugars or biomass, assumes values of around 40% and increases with spherification and even more in the carbonization to porous solids to values close to 73% (Khan et al. 2019). The reaction mechanism consistent with this sequence, for simplicity is exemplified in Figure 8 with a polymer of xylose and derivatives of lignin monomers, follows the steps already summarized in the HTC spherification, and in the preparation of furfural from hemicellulose (Chapter 2). The elimination of acid-catalyzed water in the C5 units of xylan polymers leads to the formation of sugar derivatives that can be described as α–β unsaturated ketones, together with furfural by breaking the glycosidic bond in the polymer (steps 1–2, Figure 3). Furfural and in general the formation of aromatic rings are the thermodynamically favored products. But furfuraldehydes are aldehydes, functional groups that have extensive reactivity to the level of self-condensation and cross-condensation. With the very likely decomposition of acid-catalyzed lignin into ketone substances, such as (3), an intermediate mixture of substances is generated, such as (4) or (5) containing rings with confined aromatic character. These can undergo condensation via enol on α carbonyl carbon, Diels-Alder condensations between dienes and double bonds activated by –OR groups, such as methoxide or –OH, and other reactions induced by OH radical, in initial phases induced by oxygen contact and peroxide decomposition. The originated substances have a higher molecular weight, their mixture must be liquid with increased hydrophobic properties, capable of forming spheres in the liquid or gas phase due to surface tension. The term "polymer fusion" specified in the title refers to this process where a homogeneous liquid phase is formed from the decomposition products of hemicellulose and lignin and, in part, from amorphous cellulose moieties. In the presence of oxygen, the allylic and benzylic positions are easily converted into peroxyacids, unstable species that generate free radicals, such as ·OH initiating other polymerization pathways. Radical species originate in the thermal bond breaking process are prevalent in thermal carbonizations. In this case, the same species can be formed without the need for heating as long as aerobic conditions are allowed. The partial aromatic character of (4) and (5) tends to extend resonance π pathway for adjacent carbons, the presence of free radicals provides the species capable of producing substances such as (6) and (7) shown by rearrangements in Figure 3. These substances begin to give a dark brown color to the biomass. This means that the extension of aromatization is taking place, with the fusion of the aromatic parts, probably still with the presence of oxygenated rings of pyran and furan. When the solid charge-conducting phase begins to form, a rapid increase in the rate of reactions occurs since solid-state chemistry allows for the formation of e–/ h+ pairs, the former allows for reduction, the latter catalyzes oxidative reactions on the surface of the solids and leading to their propagation to the outer limits of the solid in formation. In these stages, the evolution of gases such as CO, water and the formation of structures similar to graphene occurs, the color is now dark black and the material is a good electrical conductor (see Figure 3). For simple sugars, disaccharides or polysaccharides, equivalent reaction pathways can be written.

All reactions that lead to products similar to graphene produced in Figure 3 can be carried out at close to room temperature, if carried out in the presence of concentrated acids. The use of concentrated acids requires care and immediate contact with sugars or biomass results in violent and unspecific reactions. A way to manipulate acid reactivity must be found. The acid must have access to the entire surface of the complex fibers of the biomass in a relatively unreactive manner and be activated slowly in the subsequent phase, which suggests the use of impregnation methods. Meaning, an acid or other substance is deposited on the surface of the biomass from the diluted solution, using a very low viscosity solvent, and subsequently, the solvent is evaporated. The impregnation methods have already been the focus of attention in the degradation of biomass, namely with the use of water or diethyl ether (Meine et al. 2012). Various acids can be used, but phosphoric and especially sulfuric are particularly effective. The mixture of both is also particularly efficient.

As a solvent, water is generally not suitable, as it is very difficult to remove from biomass polymers rich in very hydrophilic sugars, producing a diluted acidic phase on the fiber surface. Diethyl ether can be used, but it has reduced solubility in water, a fact that can lead to inefficient impregnation of hydrophilic fibers found in biomass. In addition, it can react with concentrated acids to produce high molar mass decomposition product coatings in the biomass. The solvent used must be compatible with the hydrophilicity of the substrate, in order to allow good contact between the biomass voids and the solution, usually done for several hours (usually at least overnight). The solvent should also not react with concentrated acids, be non-viscous, have a low surface tension and having a low boiling point for easy removal. Methanol meets all of these parameters and is the solvent of choice because of its inability to react with concentrated acids (see

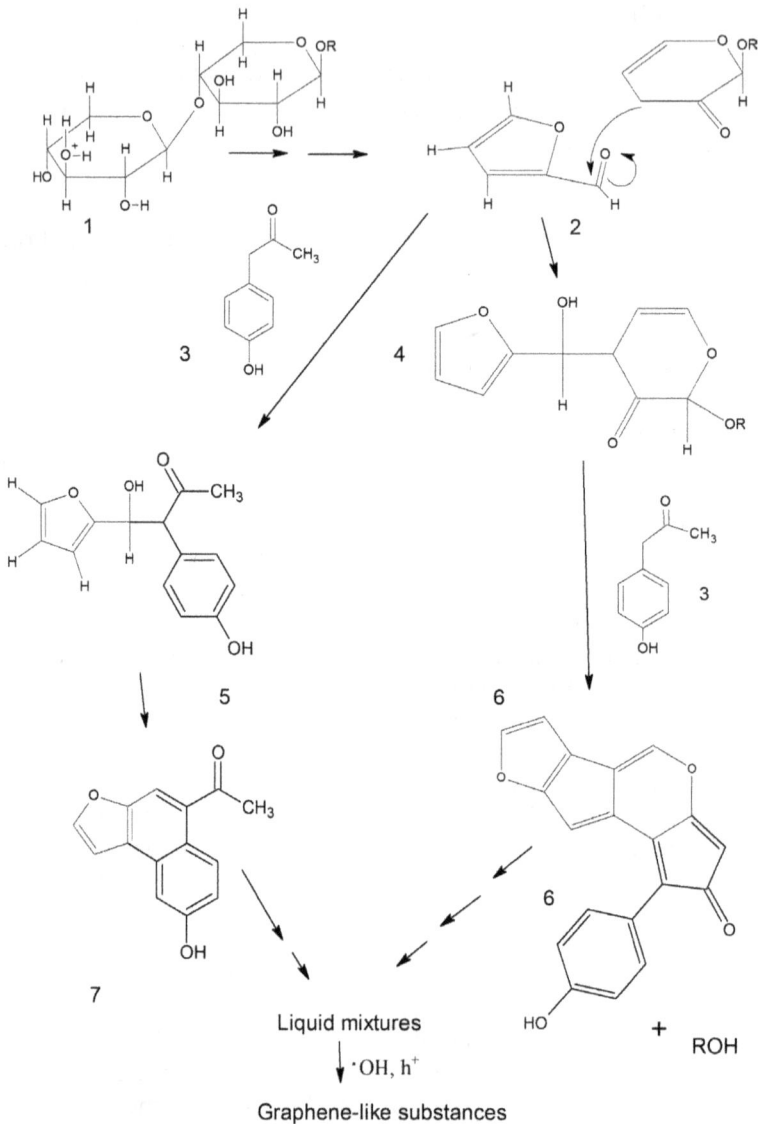

Figure 3. Reaction mechanism of the very low temperature carbonization process, VLTC. With acid-catalyzed water elimination reactions coupled to keto-enolic condensations in the presence of radical species, molecular decomposition products of both lignin and hemicellulose form a complex liquid mixture of substances, capable of forming spheres, subsequent reactions result in the formation of solid fractions with portions similar to graphene.

Chapter 6). After the removal of the solvent, the impregnated acid leads to a high local concentration on the surface, easily promoting the desired elimination reactions for water removal beginning of the biomass carbonization process, through the above-mentioned reaction pathways.

Sulfuric acid also has desiccant properties, facilitating the removal of water from sugars by elimination reactions. The mechanical movement of the impregnated biomass can assist in the process (Wu et al. 2019b) facilitating the access of the acid to the intricate structure of the biomass. Both biomass saccharification and carbonization products are possible with this methodology, depending on the processing conditions. The carbonization process can be carried out at low temperature, in the range of 50–80°C for 48 to 78 h or more, leading to extensive carbonization of the biomass. Example of conditions that allow this carbonization, at temperatures just above the environment involve dry biomass (40 to 60°C), contact with a strong acid dissolved in methanol for 4–8 h, evaporation of the solvent under vacuum and continuous mechanical stirring by a convenient method depending on the size of the biomass, at temperatures of 50–80°C, for 24–80 h. The biomass used can undergo previous milling or be used "as is", with minor hardware modifications. The method also allows biomass that is simply air-dried to be carbonized, avoiding the initial drying in an oven. Materials such as textile biomass, residual biomass and manure can be processed using this method. A great advantage of this type of carbonization is the very low energy expenditure when compared to its classic thermal techniques.

These conditions were not disclosed until 2016 and are published here for the first time. These first results have led to a better knowledge of the basic process and at the same time, its applications to different materials were tested and the product characterized. This has given the method a very attractive way of processing which adds value to biomass residues such as straw cereals or other types of biomass. This carbonization must be differentiated from the low temperature carbonization process, in which the biomass is transformed into coal by heating in a muffle at temperatures in the range of 170–400°C (Skodras and Amarantos 2004).

6.1 Very low temperature carbonization (VLTC) of wheat straw

The mechanism of polymer melting described above has as an essential mechanism the reactions of dehydration of sugar polymers catalyzed by strong acids followed by condensation steps with growth in the number of fused rings and aromatization. If the process is continued, carbon-related materials are formed. An example of using the conditions described for carbonization at low temperature is exemplified here, using wheat straw as initial biomass, including the characterization of some properties of the materials obtained. Wheat straw was previously dried at 60°C, ground in a Wiley mill with a particle size of one millimeter, and impregnated with a sulfuric acid/methanol mixture for 4 hours under stirring, followed by evaporation of the solvent under vacuum and finally carbonization at low temperature for 70 h. The final material, after washing with warm water gives rise to a very black and shiny material, samples of which are shown in Figure 4, both for the ground versions and only cut into pieces with scissors.

(a)   (b)

Figure 4. Photographs of wheat straw samples submitted to the VLTC process, ground (a) and cut into pieces (b).

The material obtained is non-flammable, with a density of 1.2–1.3, slightly brittle, stable until temperatures around 650–700°C, after which it appears to undergo a phase transition that alters its shape and properties. The determination of the elemental composition results in elementary composition values close to the expected if the xylose were converted from $C_5H_8O_5$, making two C = C double bonds, to approximately $C_5H_4O_{1.2}$. A more detailed characterization of the material obtained is presented in the following sections.

6.1.1 Elementary composition

Table 1 shows the elementary composition of different low temperature carbonization tests (VLTC) of wheat straw, with different concentrations of sulfuric acid, with a residence time of about 60 hours.

Table 1. Approximate elementary composition of samples of wheat straw subjected to low temperature carbonization with a method of impregnation with methanol of increasing sulfuric acid levels and subsequent carbonization.

$[H_2SO_4]$ μL/5g biomass	Elemental composition			Notes
	Ref.	%C	%O	%H	
0	1	42.9	38.2	5.7	Reference
215	2	43.7	38.0	6.1	
240	3	44.5	36.8	6.0	
350	4	45.3	35.3	5.9	
480	5	47.8	33.8	5.8	
600	6	47.5	33.4	5.5	
650	7	47.8	33.0	5.4	
720	8	49.5	32.5	5.3	
1500	9	56.2	32.0	5.1	

The carbon content increases with the increase in the concentration of sulfuric acid impregnated in the biomass for 60 hours of treatment with a concomitant reduction in the oxygen fraction. The final oxygen content tends to stabilize at a higher level, around 30%, when compared to the HTC process or in classic thermal carbonizations.

The evolution of the color of the sample during the process may give indications about the extension of the sequence of reactions that lead to the carbonization process. For assays that use less than 300 μL/5g biomass of sulfuric acid, the biomass does not have a black color, showing instead a dark brown color. The formation of fused aromatic rings with extended aromatic conjugation is not formed, under conditions of low sulfuric acid concentration. In treatments with a higher concentration of sulfuric acid, the resulting material shows a shiny dark black appearance, reminiscent of crystals. Wheat straw has a superficial layer of waxy materials that reaches values up to 3.2% of dry matter (Sin 2012), they are believed to act as a defense system against dehydration of the plant. Its effects on the VLTC process are not yet well known. An observation made with aged samples, kept in the dark, but in aerobic conditions (up to 1–2 years), indicates that the color is degraded to brown, in some samples treated with less acid, but that it originally had a dark black color.

6.1.2 Electrical conductivity

Electrical conductivity and color properties may indicate the degree and extent of fused aromatic ring chains, possibly due to their tendency to promote stack-stacking, the formation of graphite-related parts, possibly with a large number of defects considering the labile experimental conditions used. Both the electrical conductivity and the color suppose a large displacement of electrons of the carbon double bonds through an enlarged carbon skeleton. As the material obtained is expected to be formed by heterogeneous phases, which reflect the different initial components present in the biomass, a good electric contact between physically distinct adjacent phases should occur.

The samples obtained by VLTC carbonization were tested to verify their electrical conductivity. Both the initial biomass and the final carbonized materials show a high anisotropy in the arrangement of the carbon fibers, since the macrofibril arrangements in the wheat straw are arranged in the apical direction. Thus, axial and transverse electrical conductivity were measured. In Table 2, the average results of its determination are presented.

The results presented in Table 2 are average values of electrical resistivity of a large set of measurements series with sample aged in aerobic conditions for six months, at a temperature of 20°C. The reasons for presentation in these conditions are the stabilization of the sample. The electrical conductivity shows great variability in fresh samples, for unknown reasons, in some cases with resistance values equal to zero, these measurements are under further research. A very anisotropic difference in values is obtained, with a lower electrical resistance in the axial or apical direction in the present biomass—

Table 2. Specific resistance (ρ) of char samples obtained from wheat straw in axial and transversal directions.

Sample treatment[1]/Reference	Direction[2]	Electrical resistance	
		r (Ω.m)	r/mass (Ω.m.g^{-1})
H2SO4, 60°C, 1	A	1045	1.7
H2SO4, 60°C, 1	T	45511	455.0
H2SO4, 60°C, 2	A	1404	2.1
H2SO4, 60°C, 2	T	8889	178.3

1. Samples aged six months at 20°C in aerobic conditions. 2. A-Apical, T-transversal direction to apical.

the direction of growth of the macrofibrils, and much less in the transversal, see Table 2. Although not directly comparable, a similar effect is observed, for example, in graphite due to its sheet structure, being better conductive in the direction of the sheet and smaller in the orthogonal direction.

The electrical conductivity of the samples is also found to be stable in samples stored under an inert atmosphere, but suffers degradation for the samples left exposed to the air for prolonged periods of time.

After the high temperature heating process described above, at the temperature above 650–700°C occurs a phase change, which leads to the loss of the conductive properties of the materials obtained, showing the same values of specific resistance greater than 10E7 Ω.

6.1.3　GC-MS analysis of aqueous intermediate extracts

In an attempt to fully understand the mechanisms that occur in carbonization at very low temperature (VLTC), in the present test using wheat straw as a source of biomass, several biomass samples were taken at intervals from the beginning of the process, dissolved in water under the assistance of ultrasound and neutralized. These samples were extracted by Solid Phase Microextraction fibers (SPME) with polydimethylsiloxane/carbon fiber (PDMS/C) in an attempt to collect the largest number of substances present in solution. The fibers were subsequently injected into a GC-MS system. The objective was to monitor the formation of low/medium molar sized substances throughout the carbonization process. Figure 5 shows the structure of the identified compounds. As can be seen, hemicellulose degradation molecules are observed, such as (5) and (6), furfural and furfural alcohol, respectively; cellulose degradation, levoglucosan (1). At the same time, different fused polycyclic aromatic hydrocarbons (PAH), such as anthracene (4) and phenanthrene or one of its isomers (2), are observed. These molecules show aromatic rings fused at C6. Other chromatographic peaks appear with mass spectra that correspond to substances originated by the condensation or fusion of lignin decomposition products with C5 and/or C6 sugar decomposition substances, such as the substance (3). Several bands whose structure has not been identified, with m/z in the 190–250 g mass

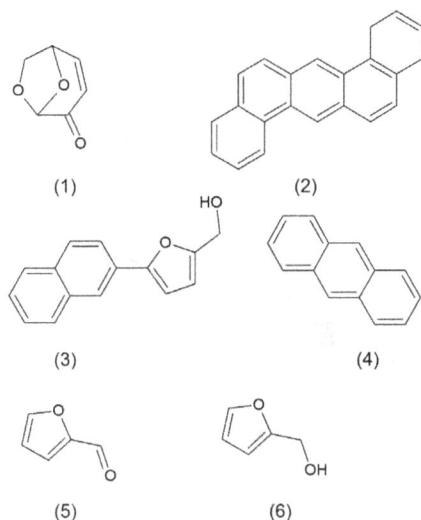

(1)   (2)

(3)   (4)

(5)   (6)

Figure 5. Substance search by SPME-GC-MS in intermediate stages of VLTC carbonization in neutralized aqueous extracts of wheat straw biomass.

range show fragmentation patterns in GC-MS-EI, which suggests their origin in condensation reactions or dimerization of decomposition products of three types of predominant substance in biomass. A proportion of uncharacterized high molecular mass molecules was also recorded.

6.1.4 *Measurement of the contact angle*

Several measurements of the contact angle of samples submitted to VLTC carbonization, with higher fixed carbon content, and stable in relation to aerobic oxidation, were made. The contact angle can provide information about the hydrophobicity or hydrophilic nature of the surface. The results give an average contact angle value equal to $90 \pm 7°$. The result is similar to that observed for graphite (Westreich et al. 2007) or carbon fiber (Tran et al. 2011).

6.1.5 *Scanning electron microscope (SEM) analysis*

SEM images can provide some insight into the materials obtained in the new type of carbonization described and samples were prepared to try to obtain images in different sets of VLTC carbonization assays. SEM images of untreated wheat straw show a structure similar to the interleaved arrangement of two types of fibers, represented schematically in Figure 6. The images show pairs of elongated apical fibrils, one consisting of arrays of cellulose macrofibrils interspersed with polymer domains of porous hemicellulose/lignin. The dimension of the hemicellulose/lignin domain is slightly larger than that of cellulose, at least for the analyzed wheat straw sample.

Figure 6. Spatial arrangement of wheat straw cell wall domains. In dark color is the super array of cellulose macrofibrils, in lighter color the porous domain of hemicellulose/lignin polymers.

SEM images of VLTC carbonization products from wheat straw are shown in Figures 7 and 8. Figure 7(a) shows an image of carbonized wheat straw biomass under the conditions mentioned in Table 1 above, experiment 8. In this, the cylindrical super bundles of cellulose macrofibrils are clearly visible, interspersed with the hemicellulose and lignin phase. In cellulose, the individual macrofibrils can be seen individually (zone 1 in Figure 7a). These appear to be in pristine shape, perhaps with the exception of the surface, and show a diameter of about 30 µm. The diameter of the super bundle or super array of macrofibrils is about 700 µm, which should indicate that each bundle of cellulose contains a number of about 450–550 macrofibrils. However, there are sites that indicate partial destruction of the cellulose macrofibrils, as can be seen in Figure 7(a), detail 4. The dimension of the hemicellulose/lignin domain is greater in relation to that of cellulose, close to 1000 µm. This domain appears to have undergone major changes (Figure 7a, zone 2) when compared to SEM images of untreated wheat straw (Zhang et al. 2015). At the site, a disorganized homogeneous phase is visible, appearing as multiple leaf-shaped formations, with a thickness equal to or less than 1.5 µm and with an irregular shape. The great presence of empty spaces, with a diameter of about 50–200 µm in width (detail 3 in Figure 7a and an amplification in Figure 7b) is visible. On the surface of the macrofiber arrays, a veiled structure with folds in diameter of 3–25 µm is observed (Figure 7a and b). The observed structure suggests the formation of a phase, not very compact, of the turbostratic graphene type (Musiol et al. 2016). The veils appear to have circular spots in some points, ranging in size from 500 nm to 1 µm (detail in Figure 7c). In some places it is also possible to observe small spherules, as seen to the right of the letter *a* in Figure 7.

An interesting feature is the arrangement of the homogeneous material "fusion polymer type" which when covering the surface of the bundles of cellulose macrofibrils does not occur in a continuous way as expected, but

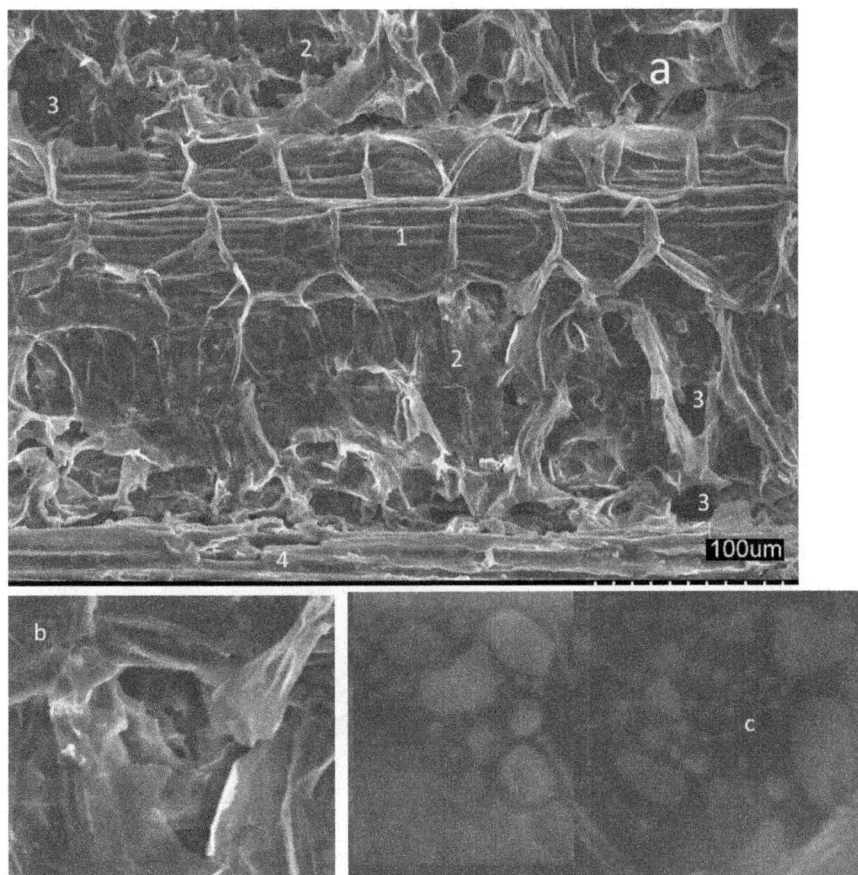

Figure 7. Details of SEM images of wheat straw in the VLTC process (see text for details).

instead shows preference to form an a regular octagonal organization in which the veils are added, forming fibers with about 40–50 μm in diameter (Figure 8a). A detail of the veil segment is identifiable in Figure 8b. This may indicate some repulsive electrostatic interaction between the veil material and the cellulose macrofibrils.

Electrostatic repulsion due to the different polarity of two phases is a possibility, however, characteristics on the surface of the macrofibril bundle, a possibility of the presence of amorphous regions at regularly separated distances may also explain the effect. If this is the case, it means that every 500 μm away in the macrofibril, its structure may change to have a different organization.

Obtaining disorganized turbostratic graphite layers at the site corresponding to the hemicellulose/lignin domain is not always observed; under the same treatment conditions, areas with differently organized domains are formed, possibly due to slightly different reaction conditions

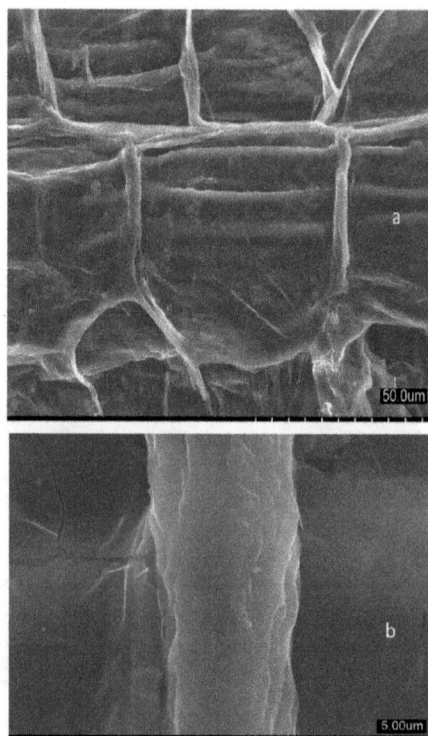

Figure 8. SEM images of carbonized wheat straw with details related to the hemicellulose/ decomposition substances on the surface of the bundles of cellulose macrofibrils.

at specific locations. In this case, an alveolar structure is formed, similar to foam, with inclusions of spherical formations on the inside, behavior shown in the images in Figure 9a. The formation of this type of material must be associated with the existence of a liquid phase, together with the simultaneous evolution of gas, a known way of making spongy materials. The gas molecules may be water or, eventually, carbon dioxide or other small molecules ejected from the decomposition of biomass, such as methanol. In Figure 9b, an amplification image of some of the alveoli that show the occurrence of different types of spherulation is shown. This observation suggests that in this phase, there was a transitory liquid medium, obtained by "fusing" the polymers where the hemicellulose/lignin phase previously existed, invoking a process similar to that described in HTC, but with the possible presence of some gas molecules. Thus, in the VLTC process, in some part of the process, even at low temperature the formation of a mixture of liquid with surface tension occurs inducing the formation of alveoles.

The observed wells have a dimension in the range of 50 to 100 μm, their walls have a thickness of 1–2 μm and appear to be made with a homogeneous material. In some it is possible to see other details on the interior walls, such as the apparent runoff of solidified liquid, usually towards the formation

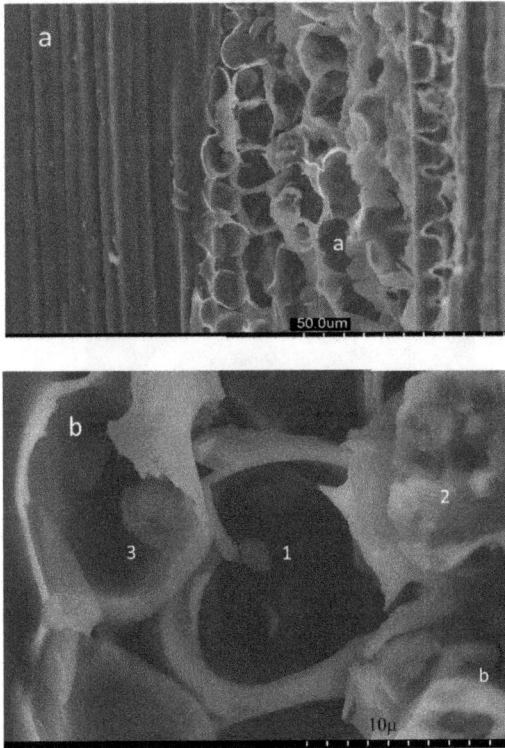

Figure 9. (a) SEM image of wheat straw samples submitted to the VLTC process showing the intense formation of a phase of an alveolar nature in the place of hemicellulose/lignin domain, (b) Internal details of the alveoli (see text for details).

of multiple spheres. The diameters of the latter can take on different values, from 5 μm to less than 500 nm (see Figure 9 details 1, 2 and 3). The mechanism of spherule formation from the liquid phase can be seen in detail in Figure 10a, which shows channels of "frozen" liquid, running down the walls of the alveolus cage (detail 1 Figure 10a) and showing at the end a large amount of spherical formation material. The double mirror aspect must mean the inversion of the position in relation to the space at the time when the process was viable and was being carried out. Their observation supports the above interpretation of intermediate formation of a liquid mixture in the decomposition of the hemicellulose/lignin domain, in processes preliminary to that of carbonization.

The bundles of cellulose macrofibrils appear to be covered with a material similar in nature to that observed in the formation of the alveoli (zone a in Figure 9a). Figure 10b shows details of an image of an alveolus at the boundary of the cellulose macrofibril bundle (zone 1, Figure 10b), indicating good compatibility with the surface material of the cellulose domain surface (details 2, Figure 10b), allowing the closure of the alveolar unit. This same alveolus shows internal spherules (detail 3 in Figure 10b).

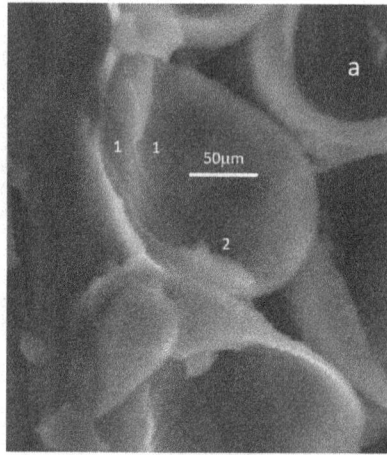

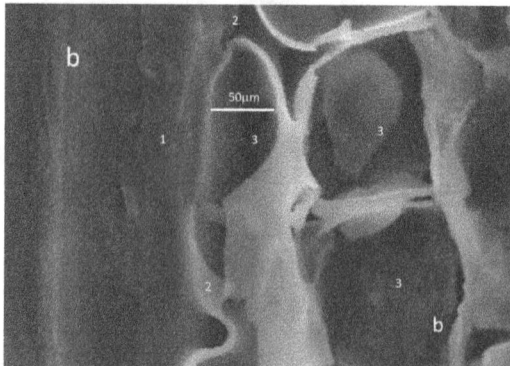

Figure 10. (a) Evidence of liquid phase movement in the wall of alveoli formed in the hemicellulose/lignin domain, (b) Details of an alveolus at the interface with the cellulose macrofibril bundle phase.

6.1.6 *Interpretation of observations*

From the analytical characterizations made in the carbonization of wheat straw by the VLTC technique, a mechanistic interpretation of the process that occurs is elaborated. Due to the composite nature of biomass and the reactivity of the main three components, namely hemicellulose, cellulose and lignin, with respect to acids, it is expected that the acid will start to attack hemicellulose. The reactions will certainly be similar to those seen in the HTC case, discussed above.

The process begins with the elimination of water in the sugars, with the formation of double bonds that rearrange to aldehyde or ketones. Along with this, the close lignin undergoes acid-catalyzed depolymerization starting to produce phenolic monomers capable of condensing with ketones and furfurals. At some point, a complex mixture of substances is formed, in the form of a liquid with a high surface tension, which is capable of forming spheres. High surface tension means strong intermolecular forces or the formation of two phases of liquids with different types of polarity. The release of molecules in the gas phase, in the presence of that liquid phase, originates the formation of the alveolar structure seen in some images that refer to the hemicellulose/lignin domain. In the absence of gas or possibly due to a different geometric arrangement, the sequence of processes leads instead to the formation of veil-like structures. Evaluating the finding of polycyclic aromatic compounds (PAH) and aldehydes that are detected in aqueous extracts from the intermediate phases of the carbonization process, these substances must be involved in the formation of the intermediate liquid phase. The final product has typical carbonization characteristics, such as black color and electrical conductivity, and thermal stability, so in the final stages the formation of conjugated aromatic rings over an extensive carbon skeleton is likely. Possibly the formation of domains similar to graphene sheets—capable of π-stacking originating domains of graphite or turbostratic graphite is supported by the observations seen. The final hydrophobicity of the products is supported by measurements of the contact angle with water, with values close to those obtained for graphite and carbon fiber. In the space where the hemicellulose/lignin domain should occupy, it shows the presence of innumerable empty spaces, probably due to its porous nature, related to the water/nutrient conducting vessels originally present.

On the other hand, it is important to understand the reason why relatively high levels of oxygen are found in the elemental composition of the obtained carbonization products. Apparently, in the mild conditions used in VLTC, carbonization does not affect crystalline cellulose to a great extent, beyond the outer surface. This is one of the major differences in relation to materials such as carbon fiber, where the fixed carbon content is greater than 90%, a fact shared with some types of coal prepared by the thermal process. SEM images also suggest that the majority of macrofibril bundles are apparently intact, although it is possible to see, in some samples, the partial destruction of their structure. On the surface, clear changes occur, since either the bundles or the cellulose macrofibrils appear to be covered by a veil, in some places even forming the casual spherification, see Figure 11. The cellulose macrofibrils appear to have a polymer coating of low DP xylan, which can easily form the liquid "polymer melting" phase discussed above, responsible for spherulation. Simulation studies of elementary composition, considering the hemicellulose/lignin domain target of carbonization, keeping the cellulose content intact, result in values for the elemental composition close to that verified, suggesting that, at least in wheat straw, the VLTC carbonization does not significantly affect the crystalline cellulose domain.

The electrical conductivity is higher in the apical direction, suggesting that phases with fused aromatic rings, similar to graphene, should constitute the lining of the bundles of cellulose macrofibrils, and that they are responsible for good electrical conductivity. Very low values for electrical resistance

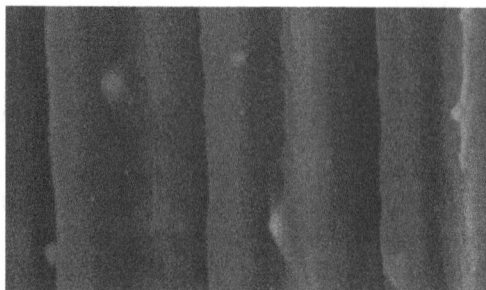

Figure 11. SEM image of the surface of bundles of wheat straw macrofibrils, covered with a polymer veil and with sparse random presence of spherules.

are sometimes measured, especially in fresh samples, this phenomenon being investigated. In the transverse direction, the structure forms random direction veils sometimes, other times alveoli, clearly making the path to the electrons π more difficult, probably requiring physical and electrical contact with different phases in which there is the presence of a greater number of imperfections in the structures similar to graphene, which may explain the lower electrical conductivity observed.

The samples submitted to VLTC carbonization conditions with lower concentrations of sulfuric acid result in carbonized materials that degrade over time, under storage conditions such as room temperature and in aerobic conditions, to give rise to dark brown materials and with loss of electrical conductivity. The interpretation is the existence of a high number of imperfections in the construction of the fused aromatic rings, most likely with the existence of multiple benzylic and/or allylic carbons in the π electron delocalization pathway. These are easily oxidized by atmospheric oxygen at a temperature as low as 15°C, leading to loss of π conjugation in specific points of carbon skeletons. The mechanism to explain this effect is proposed in Figure 12. The allylic or benzylic positions easily add atmospheric oxygen to produce peroxides, these are unstable and rearrange by adding oxygen to the double bonds, leading to their break. The same were involved in conducting electrons π, leading to the fading of the color to brown and the degradation of electrical conductivity. The eventual transport of holes (h+) to these positions followed by oxidation of water to generate reactive oxygen species (ROS) may assist the process, meaning that the process may not require oxygen.

It should be emphasized that these results refer to VLTC of the wheat straw biomass, in which the arrangement of cellulose macrofibrils is parallel and particularly strong in the axial direction. The results may be different in types of biomass that tend to cross the macrofibrils in a less compact arrangement as in other types of biomass with a low cellulose content. In any case, carbonization is always achieved.

The macrofibrils apparently untouched in some samples of materials obtained by this process can have the potential for the preparation of composite materials, since they combine the excellent mechanical properties

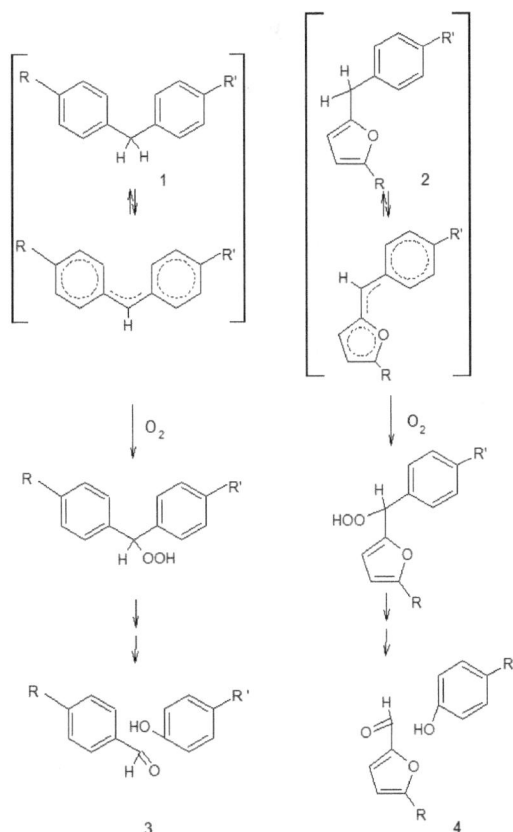

Figure 12. A possible mechanistic explanation for the color fading and loss of electrical conductivity in some of the wheat straw carbon samples prepared by means of VLTC, under prolonged storage in the air.

of microcrystalline cellulose together with a modified external environment, a more hydrophobic domain compared to the earlier existant hemicellulose/ lignin, allowing greater compatibility with the epoxy resin generally used in the manufacture of composites (see below).

The biomass of wheat straw in unground form or cross-organized like textiles, after compression can also be subjected to VLTC carbonization, making the products obtained desirable for obtaining high temperature insulating materials, ablative materials and also for construction of materials similar in appearance to carbon fiber, for ornamentation purposes, or floor coverings.

7. Preparation of carbonized wheat straw/epoxy resin composites and study of their mechanical properties

In the perspective of studying the mechanical properties of the carbonization products of wheat straw biomass catalyzed by low temperature sulfuric acid

(VLTC), one of the main issues raised was the determination of the mechanical properties of the materials obtained. The determination of such parameters is far from easy, and requires equipment that is not always available, so the use of indirect methods may provide data on these properties. It is believed that some types of commercial carbon fiber with a high modulus of tensile strength and elasticity (Young's modulus), superior to steel, are originated in carbon fiber by a turbostratic structure related to graphite. The carbon materials of wheat straw prepared in the present work seem to have phases similar to turbostratic graphite together with crystalline cellulose fibers in a relatively intact condition. It was already known that MCC and CNC's have a high potential to be used in the preparation of high mechanical strength composites. Crystalline cellulose has comparable to superior mechanical properties relatively to carbon fiber, and only smaller than that of multi-walled nanotubes, with tensile strength values of 7.5 GPa and Young's Modulus up to 20 GPa (Ghasemi et al. 2018). One of the biggest challenges in an introduction of nanocellulose crystals in composite formulations is the high polarity surface of the MCC or CNC which makes it difficult to adhere and contact with epoxy resins (Gan et al. 2017, Wu et al. 2018a). The contact angle obtained in the carbonization samples of wheat straw suggests a very hydrophobic surface, which eventually may facilitate compatibility with epoxy resins and the easier preparation of composite material.

This led to the verification of the possibility of preparing composite materials with VLTC wheat straw and to the study of their mechanical properties. These were made with commercial epoxy resin in the proportion of 40/60 (w/w, fiber/resin). The appropriate form was provided to the composites in order to allow the determination of their mechanical properties in an extensiometer. For comparison purposes, wheat straw composites, carbon fiber filaments and only epoxy resin were also prepared. Some of the results obtained are obtained in Table 3.

Table 3. Mechanical parameters of different composites made of epoxy resin and carbon materials.

Sample	Material[1]	Rupture tension[2] (MPa)	Young modulus (GPa)	Young M. comparison to CF (%)	Notes
1	WSC/ER	7.32	1.80	18.6	
2	WSC/ER	7.54	1.88	19.5	vacuum
3	WS/ER	8.73	1.75	18.1	
4	ER	32.85	2.35	24.3	reference
5	CF/ER 1	---- 3	9.23	–	vacuum
6	CF/ER 2	---- 3	10.10	–	

1. WSC—VLTC carbonized wheat straw fiber, ER—epoxy resin, WS—dried, not carbonized wheat straw fiber, CF—commercial carbon fiber filaments; a proportion of epoxy resin fixed at 60% (by weight). 2. When possible to obtain. 3. High values prevented its precise determination.

Difficulties were encountered in preparing the composites due to air bubbles trapped inside the carbonized biomass fibers. Thus, the tensile strength parameter often does not reflect the potential of the materials, and should be considered carefully. However, better information can be collected with the data related to the parameter modulus of elasticity. The results in Table 3 indicate that composites prepared with the incorporation of carbonized wheat straw fiber have lower tensile strength when compared to simple epoxy resin specimens. Carbon fiber composites are especially resistant. Comparing the elasticity module, the use of carbonized wheat straw fibers, under the same conditions of preparation and testing, results in the obtaining of values in the order of about 20% Young modulus of that obtained with carbon fiber, comparison of samples 1–3 with 5–6 in Table 3. Analysis of the broken samples revealed problems of contact between the carbonized straw fibers with the epoxy resin. In this area, research can be carried out using different epoxy resins or with processes that improve the contact between the two phases. However, a slight improvement in the mechanical properties of the carbonized fiber was observed in relation to the dry ground wheat straw, verified in the tests 1–2 in comparison with 3. More studies are necessary to improve the techniques involved, but these results show the opportunity to use cereal straw as a fiber with good properties for the construction of epoxy resin composites, for a wide range of uses. The properties of the collected materials can transform straw and other cereals, needles that come out of various trees, in interesting materials for the more economical manufacture of composites with properties superior to synthetic versions and with new properties.

8. References

Baseri, J. R., Palanisamy, P. N. and P. Sivakumar. 2012. Preparation and characterization of activated carbon from Thevetia peruviana for the removal of dyes from textile waste water. Advances in Applied Science Research 3: 377–383.

Binder, C., Bendo, T. and G. Hammes. 2017. Structure and properties of *in situ*-generated two-dimensional turbostratic graphite nodules. Carbon 124: 685–692.

Byrne, N., De Silva, R., Ma, Y. B., Sixta, H. and M. Hummel. 2018. Enhanced stabilization of cellulose-lignin hybrid filaments for carbon fiber production. Cellulose 25: 723–733.

Cao, J., Zhao, W. and S. Gao. 2018. Properties and structure of *in situ* transformed pan-based carbon fibers. Materials 11: 1017.

Ferreira, L. C. C. 2017. Study of structure of asphaltenes using spectroscopy. PhD Thesis. Rio de Janeiro (Brasil) University.

Gan, L., Liao, J. L., Lin, N., Hu, C. L., Wang, H. L. and J. Huang. 2017. Focus on gradient wise control of the surface acetylation of cellulose nanocrystals to optimize mechanical reinforcement for hydrophobic polyester-based nanocomposites. ACS Omega 2: 4725–4736.

Gao, Z. M., Jin, H. Z., Li, X. S. and Z. Hua. 2003. Phase transformation mechanism of graphite-turbostratic graphite in the course of mechanical grinding. Research in Chinese Universities Chemical 19: 216–218.

Ghasemi, S., Behrooz, R., Ghasemi, I., Yassar, R. S. and F. Long. 2018. Development of nanocellulose-reinforced PLA nanocomposite by using maleated PLA (PLA-g-MA). Journal of Thermoplastic Composite Materials 31: 1090–1101.

Huang, Y. and G. Zhao. 2016. Preparation and characterization of activated carbon fibers from liquefied wood by KOH activation. Holzforschung 70: 195–202.

Hussain, A., Mehdi, S. M. and N. Abbas. 2020. Synthesis of graphene from solid carbon sources: A focused review. Materials Chemistry and Physics 248: 122924.

Jawad, A. H., Razuan, R., Appaturi, J. N. and L. D. Wilson. 2019. Adsorption and mechanism study for methylene blue dye removal with carbonized watermelon (Citrullus lanatus) rind prepared via one-step liquid phase H2SO4 activation. Surfaces and Interfaces 16: 76–84.

Jiang, X. F., Ouyang, Q., Liu, D. P., Huang, J., Ma, H. B., Chen, Y. S., Wang, X. F. and W. Sun. 2018. Preparation of low-cost carbon fiber precursors from blends of wheat straw lignin and commercia textile-grade polyacrylonitrile (PAN). Holzforschung 72: 727–734.

Khalid, K., Asimi, A., Ahmadasimi, A., Ahmad, K., Tau, L., Yong, Tau L. and Y. Yong. 2016. Lignin extraction from cellulosic biomass using sub- and supercritical fluid technology as precursor for carbon fiver production. Journal of Japan Institute of Energy 96: 255–260.

Khan, T. A., Saud, A. S., Jamari, S. S., Ab Rahim, M. H., Park, J. W. and H. J. Kim. 2019. Hydrothermal carbonization of lignocellulosic biomass for carbon rich material preparation: A review. Biomass & Bioenergy 130: 105384.

Kubo, S., Uraki, Y. and Y. Sano. 2003. Catalytic graphitization of hardwood acetic acid lignin with nickel acetate. Journal of Wood Science 49: 188–192.

Lau, K. T. and D. Hui. 2002. The revolutionary creation of new advanced materials—carbon nanotube composites. Composites Part B-Engineering 33: 263–277.

Liu, Y. and S. Kumar. 2014. Polymer/carbon nanotube nano composite fibers—A review. ACS Applied Materials & Interfaces 6: 6069–6087.

Luo, Y. P., Li, Z., Li, X. L., Liu, X. F., Fan, J. J., Clark, J. H. and C. W. Hu. 2019. The production of furfural directly from hemicellulose in lignocellulosic biomass: A review. Catalysis Today 319: 14–24.

Meine, N., Rinaldi, R. and F. Schueth. 2012 Solvent-free catalytic depolymerization of cellulose to water-soluble oligosaccharides. Chemsuschem 5: 1449–1454.

Mirkouei, A., Mirzaie, P., Haapala, K. R., Sessions, J. and G. S. Murthy. 2016. Reducing the cost and environmental impact of integrated fixed and mobile bio-oil refinery supply chains. Journal of Cleaner Production 113: 495–507.

Mochidzuki, K., Soutric, F., Tadokoro, K., Antal, M. J., Toth, M., Zelei, B. and G. Varhegyi. 2003. Electrical and physical properties of carbonized charcoals. Industrial & Engineering Chemistry Research 42: 5140–5151.

Musiol, P., Szatkowski, P. and M. Gubernat. 2016. Comparative study of the structure and microstructure of PAN-based nano- and micro-carbon fibers. Ceramics International 42: 11603–11610.

Romero-Anaya, A. J., Ouzzine, M., Lillo-Rodenas, M. A. and A. Linares-Solano. 2014. Spherical carbons: Synthesis, characterization and activation processes. Carbon 68: 296–307.

Rubel, R. I., Ali, Hasan Md. and J. Abu. 2019. Carbon nanotubes agglomeration in reinforced composites: A review. Aims Materials Science 6:756–780.

Sin, E. H. K. 2012. The extraction and fractionation of waxes from biomass. Phd Thesis, University of York, UK.

Skodras, Panayiotis Mr. and S. Amarantos. 2004. Overview of Low Temperature carbonisation: Present Status—Properties, Yields and Utilisation of LTC chars—Survey of Various Methods—Pre-treatment Conditions & Effects—Advantages, Economic & Technological Development. CE.R.T.H./I.S.F.T.A. Report prepared in the framework of CFF OPET–E.C. co-funded project. Contract # NNE5–2002–97, WorkPackage # 4: Promotion of Low-Temperature Carbonisation (LTC) Technology. Scientific Coordinator Dr. George Skodras.

Tran, L. Q. N., Fuentes, C., Dupont-Gillain, C., Van Vuure, A. and I. Verpoest. 2011. Wetting analysis and surface characterisation of coir fibres used as reinforcement for composites. Colloids and Surfaces A: Physicochemical and Engineering Aspects 377: 251–260.

Wang, L., Skreiberg, O., Van Wesenbeeck, S., Gronli, M. and M. J. Antal. 2016. Experimental study on charcoal production from woody biomass. Energy & Fuels 30: 7994–8008.

Westreich, P., Fortier, H., Flynn, S., Foster, S. and J. R. Dahn. 2007. Exclusion of salt solutions from activated carbon pores and the relationship to contact angle on grafite. Journal of Physical Chemistry C 111: 3680–3684.

Wu, Z., Xu, J., Gong, J., Li, J. and M. Lihuan. 2018a. Preparation, characterization and acetylation of cellulose nanocrystal allomorphs. Cellulose 25: 4905–4918.

Wu, J., Chandra, R. and J. Saddler. 2019b. Alkali-oxygen treatment prior to the mechanical pulping of hardwood enhances enzymatic hydrolysis and carbohydrate recovery through selective lignin modification. Sustainable Energy & Fuels 3: 227–236.

Xuefeng, J., Ouyang, Q. and D. Liu. 2018. Preparation of low-cost carbon fiber precursors from blends of wheat straw lignin and commercia textile-grade polyacrylonitrile (PAN). Holzforschung 72: 727–734.

Yusofab, N. and A. F. Ismailab. 2017. Post spinning and pyrolysis processes of polyacrylonitrile (PAN)-based carbon fiber and activated carbon fiber: A review. Journal of Analytical and Applied Pyrolysis 93: 1–13.

Zhang, H., He, C. X., Yu, M. and J. J. Fu. 2015. Texture feature extraction and classification of SEM images of wheat straw/polypropylene composites in accelerated aging test. Advances in Materials Science and Engineering 2015: 397845.

Zhang, J. T., An, Y., Borrion, A., He, W. Z., Wang, N., Chen, Y. R. and G. M. Li. 2018. Process characteristics for microwave assisted hydrothermal carbonization of celulose. Bioresource Technology 259: 91–98.

Zheng, X. Z., Chen, M. F., Ma, Y. S., Dong, X. P., Xi, F. N. and J. Y. Liu. 2017. Enhanced electrochemical performance of straw-based porous carbon fibers for supercapacitor. Journal of Solid State Electrochemistry 21: 3449–3458.

Chapter **6**

The Methanol/Sulfuric Acid System for Cellulose Saccharification

1. Introduction

At the beginning of this century, the conversion of residual biomass from agricultural activity into ethanol became a greatly desired goal, unfortunately for everyone, the goal has yet not been achieved. The purpose turned out to be much more difficult than expected. It is possible, Nature shows it, for example in the digestive system of the termite, it can digest various types of wood in about six hours, releasing as a residue essentially lignin (Griffiths et al. 2013). More extraordinary, the termite makes the complete metabolism of the sugars present in the biomass in a gastric system similar to a tube of approximately 6 × 1 mm, in length and diameter respectively. It is aided in the process by a large number of symbiotic microorganisms that cooperate in the process of decomposition of sugar polymers, exchanging substances with the host in mutual benefit. Analytical techniques capable of characterizing communities of microorganisms are not yet well developed, and several fields of human activity could improve from a greater availability of such techniques. Biomimetic saccharification is a desirable goal, as it will enable the production of a different number of substances derived from sugars with processing technology developed during a long evolutionary process. Thus, the focus on research activity on polysaccharide saccharification can be assessed by the large number of publications on the subject in the past 20 years. Unfortunately, many of them do not take into account the molecular structure of biomass and result in only very low ethanol yield, after which no competitive conversion is viable.

Any attempt to de-polymerize the polysaccharides present in the biomass, which reach levels up to 80% of the dry mass, faces some problems. The main ones are:

(1) The only biomass component with good accessibility is hemicellulose, a polymer predominantly made up of C5 sugars, a fraction that can be recovered almost entirely in the form of sugar (≈ 30%), along with a small fraction of hexoses.

(2) In many types of biomass, cellulose is difficult to access because it forms or is inserted in a complex mesh, and it is present in crystalline form, especially reinforced with networks of different types of chemical bonds, making it difficult to mobilize by enzymatic exfoliation techniques or prolonged chemical or other de-crystallization technique.

(3) The use of enzymes in the degradation of large crystals, no matter how active they are, makes their immobilization impossible, which in turn makes their reuse not possible. Enzymes are expensive and their single use is unthinkable.

(4) Organisms capable of converting sugars in high concentration to ethanol allow its isolation by simple processes, but they are scarce and only capable of fermenting C6 sugars, as known of now.

These problems are being subjected to intense research however, the truth is that the process of converting biomass into cellulosic ethanol is not implemented beyond the demonstration unit, and the ethanol yield and associate price are not competitive. In any case, the residual biomass must not be burned for any purpose. As previously demonstrated high value bioproducts can be obtained from biomass. The logical approach is to take advantage of each fraction of biomass, with the processes that are currently available, as new technologies emerge and that eventually manage to achieve the goal of cellulosic ethanol production. In this work, thermo-chemical biomass treatments were defined from the outset. However, it must be said that they may have the potential, together with the treatment of other residues, to complement the desirable preparation of alcohols derived from biomass.

2. Saccharification of biomass

To obtain ethanol from biomass, it is necessary to first saccharify it. Based on its structure, a sequential process seems adjusted, whether the biomass is treated or not by physical processes. The easiest fraction to be removed by suitable treatments is hemicellulose, followed by the amorphous part of the cellulose, the remaining material essentially consists of crystalline cellulose with a little lignin. Conditions that lead to the eventual recovery of glucose contained in the crystalline cellulose part have been the most difficult part to perform. If done enzymatically, a cocktail of cellulases, of the endo and exo type together with glycosidases should be provided. Nature optimizes and manages to do the process, expressing and repressing multiple enzymes, sequentially and yielding a relatively low sugar concentration, but suitable for biological systems. The sequence of these processes is not yet possible to be done artificially, and the same process must be modified, as the human interest is the production of sugar solutions in high concentration. Attempts to approximate the described sequence have been studied, and are in a primordial phase, being studied in the Consolidated Biomass Processing (CBP) where the natural strategy is replicated using communities of microorganisms.

If the same depolymerization sequence of the biomass constituents is carried out by chemical means, the conditions necessary for the removal of the pentoses do not interfere with the cellulose polymers and those necessary for saccharification of the hexoses degrade the already hydrolyzed pentose monomers. This again points to a sequential process, which is suggested in this book for the concept of biorefinery.

3. The capacity of the sulfate anion to de-crystallize/hydrolyze cellulose at low concentration in water at high temperature

Tests of sulfate anions in reaction medium with heating/catalysis with microwave irradiation in slightly alkaline water, at temperatures up to 230°C, shows the formation of oligosaccharides from crystalline cellulose. Strict control of the experimental conditions must be exercised, since the final products depend on small variations in pH, type of base, temperature and reaction time. The presence of acids tends to produce products similar to those obtained in the carbonization process in aqueous medium (HTC), described in the previous chapter (Reza et al. 2015). Water purified by double distillation, passed through mixed ion-exchange resin treated with activated carbon with electrical resistance greater than 18 MΩ is also able to promote the carbonization of neutral cellulose fiber. The effect is much more pronounced when using various types of dry biomass. The explanation lies in the trace presence of hemicellulose, rich in acetyl groups (CH_3CO), which in rapid hydrolysis provides the acid capable of initiating the carbonization process by a process similar to that obtained in HTC. When using paper pulp in pure water, conditions of low acid concentration from biomass give rise to the brown color characteristic of condensation products of glucose decomposition substance, along with some degree of carbonization.

Hydrolysis of cellulose assisted by sulphate ions in a slightly basic medium gives rise to oligosaccharides and, under certain conditions, to cellulose crystals of colloidal size, capable of gelling. In Figure 1, 1) a HPLC chromatogram of a cellulose fiber sample treated with water at 210°C is presented, under microwave radiation in the presence of 1% (w/v) calcium sulfate ($CaSO_4$), in slightly alkaline conditions. The products obtained are a mixture of glucose (a), cellobiose (b) and mixtures of oligosaccharides, one centered on about seven glucose units (c) and mainly a mixture of others with between 22 and 24 glucose units. The size of the oligosaccharides was inferred by comparing the HPLC retention time of the corresponding oligosaccharides from a sample of corn syrups. Tests carried out under the same conditions, but in less alkaline conditions, such as a buffer solution of 1% (w/v) sulfuric acid and 1% (w/v) calcium sulfate, produce a drastic change in the final products of reaction, with the main formation of glucose and a small amount of oligosaccharides with 4–6 glucose units, a result shown in Figures 1, 2). With the same substrate in alkaline conditions, but using even higher temperatures, in the range of 230–240°C, colloidal dispersions

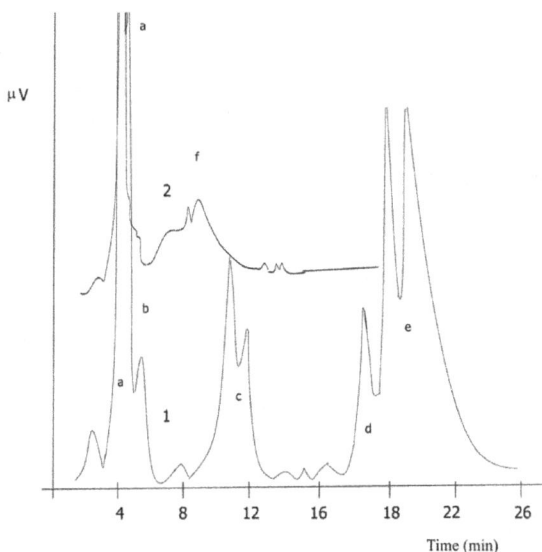

Figure 1. HPLC chromatograms of pulp paper hydrolysates with water at high temperature (210°C) with microwave heating: (1) in the presence of 1% (w/v) calcium sulfate (CaSO4) producing glucose (a), cellobiose (b), oligosaccharides centered on approximately 7 glucose units (c) and oligosaccharides centered on the app. to 22–24 glucose units (d) and (e); (2) in the presence of 1% (w/v) sulfuric acid and 1% (w/v) calcium sulfate (CaSO4) under the same conditions, producing mainly glucose and oligosaccharides with 4–6 glucose monomers (f).

are obtained by cooling, in the form of a transparent gel, meaning that a crystalline cellulose structure has not been completely destroyed in previous processes. In this case the surface appears to be hydrolytically exfoliated to the surface to give rise to particles of colloidal size, at least under the aforementioned conditions.

The results described above suggest that sulfate anion and sulfuric acid are especially effective in interfering with crystalline cellulose fibers, catalyzing their de-crystallization/hydrolysis.

4. Biomass saccharification with strong acids

It has long been known that strong acids such as phosphoric and especially sulfuric acids are capable of promoting the saccharification of biomass and cellulose (Moe et al. 2012). In fact, claims of using them under viable conditions to convert biomass to ethanol are found in the literature (Sun et al. 2011, Harmer et al. 2009). The use of concentrated sulfuric acid at low temperature is known to saccharify biomass in high yield with a little formation of degradation products, namely furfural that are strong fermentation inhibitors. This means that the sulfuric acid must be able to de-crystallize the cellulose at the same time that it promotes the hydrolysis of the fibers obtained in oligosaccharides and, subsequently, in sugars. Fortunately, glucose is stable in cold solutions of concentrated sulfuric acid, allowing it to recover by rapid dilution, for example. The main problems attributed to saccharification of biomass with the use of strong concentrated acid are the presence of pentoses, which are not stable in their presence at room temperature or even at low temperatures, leading to the formation of decomposition products. Although xylose is stable in concentrated acid at temperatures close to 0°C, under conditions of higher temperature, such as 20°C, it begins to be converted into furfural and tends to polymerize. This reason, which is associated with the difficulty in fermenting the pentoses, leads to the fact that from the biomass hydrolyzate the sugars in C5 must be isolated, or to proceed to the previous removal of the hemicellulose fraction. The fermentation of the recovered pentoses must follow a different route, or different destination, as discussed in Chapter 2. Another problem found in the successful saccharification of biomass is getting at the end of a mixture of sugar/concentrated acid. Sulfuric acid is one of the cheapest chemical reagents, but its recovery is necessary to make the process competitive. In the literature, techniques such as countercurrent ion exchange chromatography (Wooley et al. 1998) or single membrane electrodialysis-induced separation (Palaty and Zakova 1996) are considered to be capable of separating the resulting acid/sugar mixture. Although these separation methods are effective, the separation processes are complicated and expensive to maintain. Different techniques must be developed to allow the production of ethanol in a more accessible way from the biomass hexoses.

In any of the above processes, obtaining sugar from cellulose presupposes two different processes: (i) first, the conversion of crystalline cellulose into a more amorphous material; (ii) second, the hydrolysis of the $\beta(1,4)$ glycosidic bonds to obtain oligosaccharides and, in more advanced stages, monosaccharides.

Sulfuric acid at temperatures below 10°C, in concentrations of about 70%, is capable of solubilizing biomass, in some cases with the assistance of phosphoric acid. In this process, crystalline cellulose is somehow either converted to amorphous form, or defibrillated, versions more conducive to its hydrolysis. A great deal of research has been done in the last few years of solubilizing biomass using ionic liquids that are also capable of achieving the same solubilization (Zhang et al. 2014). However, these substances are expensive and potentially toxic, preferring mixtures of sulfuric and phosphoric acid that achieve similar results. The use of phosphoric and/ or sulfuric acids also has the advantage of catalyzing the hydrolysis of the glycosidic bond and, eventually, serving as a nutrient in subsequent fermentation medium, if ethanol production is the ultimate goal.

4.1 Methods of acid impregnation

More recently, different techniques for the use of acids have been tested with the potential to improve the results of biomass saccharification. One of them is the strong acid impregnation technique with the aid of solvents and mechanical means (Schneider et al. 2017). The impregnation of biomass with acid has been known for some time, but it is usually done in aqueous solution (Bakker et al. 2004). The use of water gives rise to several problems, leading to greater difficulty in removing it from the very hydrophilic polymers in biomass. Impregnation methods require less biomass preparation, sometimes avoiding grinding, and a simple air drying process is sufficient to precede the acid impregnation technique. A major breakthrough was the use of organic solvents, the concentrated sulfuric acid can for example be dissolved in ethyl ether (Meine et al. 2012), followed by the removal of the solvent to leave the acid in the hydration water layer of the carbohydrate polymers. If assisted concomitantly with mechanical movements, the impregnated biomass undergoes decrystallization/hydrolysis with good efficiency. These methods have generally allowed better results in the attempt to convert biomass into ethanol than those carried out in previous research phases, even with the use of enzymes, and continue to be the subject of detailed research.

5. Mechanism of decrystallization of cellulose with strong acid

The reasons why sulfuric acid is very efficient in promoting the decrystallization of cellulose are not very clear. The involvement of different cellulose allomorphs is invoked by analyzing an old process, the mercerization of cotton. In this process, the hydrogen bonds of the hydroxyl groups involved in the formation of the crystalline network are altered in terms of the structure of the network, leading to visible effects such as swelling of the fibers, with possible rearrangement of the position of the saccharide monomers, and sometimes invoked a reversal of the direction of the glucose chains. The change in the hydrogen bonding networks gives rise to the predominance of the type II allomorph. Allomorphs with a less dense molecular arrangement may explain the observed expansion effect, which must be associated with some degree of decrystallization.

The hydrogen ion (H+) must diffuse very easily through the crystalline network, taking into account its size and the nature of the crystal's hydrogen bonding network. This diffusion will create a surface potential that induces the entry of counter ions, as long as they have the requirements to enter. In this case, either the sulphate or the phosphate is able to pass through the layers of the crystalline network. Sulfate is especially efficient and must have adequate structural characteristics to enter the network, despite its relatively large size. Models of cellulose allomorph I show the intramolecular 2-6 or intermolecular 2-6′ hydrogen bond, both of which are likely to interfere with the sulfate, allowing it to enter (see Figure 2c). At first glance, it could be expected that acids with small counter ions such as F– or Cl– would have easier entry, but the truth is that the sulfate or phosphate anions appear to be more efficient. The fact that they have four oxygen atoms in a tetrahedral position, capable of forming hydrogen bonds at different angles, with the simple rearrangement of H+, allows the substitution of hydrogen bonds with suitable angles and are available in different positions to make the

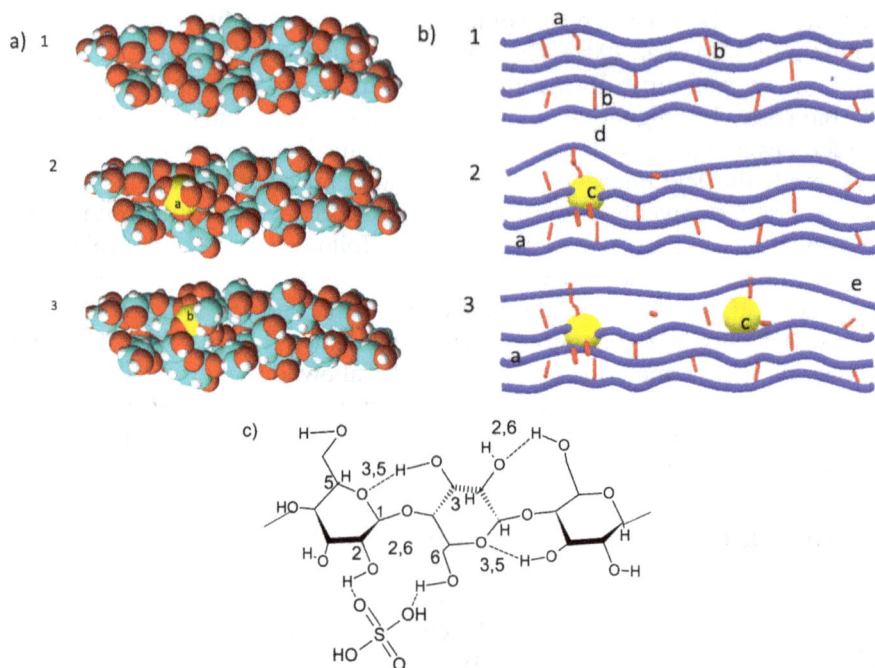

Figure 2. (a) A pair of D-β-(1,4)-octasaccharide with minimized energy as a model for cellulose crystal network, 1. A sulfate anion (a) is placed successively near the interface between the fibrils, gradually replacing several hydrogen bonds with a pair of two involving sulfate oxygen atoms (see c). A summary sequence is shown in (a) as 1-2-3, in the last image the sulfate anions appear to be positioned inside the lattice with minor conformational changes in the glucose units. The joint effect of the entry into the crystalline network of several sulfate anions is illustrated, which begin to defibrillate the simple cellulose chains as loose fibrils, already disconnected from crystal lattice. (b) The same effect with a simpler visualization; the blue lines represents single cellulose fibers, red lines interfibrils hydrogen bonds and at yellow the incoming sulphate anion, with the same overall final effect. (c) Initial single fibril 2-6 hydrogen bond break by substitution with a pair of hydrogen bonds from sulfate oxygens.

connections, with a minimum global energy variation of the system. This availability of four connection points with the crystalline network, capable of multiple interactions, means that low energy configurations are found, by simple vibration, which seems to be an essential structural characteristic. This allows diffusion of the sulfate or phosphate anions without breaking the hydrogen bonds of the crystalline network, but instead replaces the bond with others, such as a pair of hydrogen bonds, using two or more oxygen atoms from the sulfate anion. This substitution of hydrogen bonding may allow a gradual entry of sulfate anion into the crystal with conformational adjustments in the glucose bonding networks.

Figure 2 shows an image referring to computational models of the microfibril contact zones of the cellulose allomorphs I interface, showing the normal hydrogen bonds between the 2-6' glucose carbons (see Chapter 1). Sulfuric acid and phosphoric acid, due to their geometric shape, can replace this hydrogen bond with a small variation of global energy. Its entrance can be easily visualized by replacing a hydrogen bond with a pair of the same with the oxygen atoms available in the sulfate anion, leading to its placement in the layer between two cellulose microfibrils. This positioning and the repetition of the same process with more interior hydrogen bonds, such as one 3-5′ bond, will allow the sulphate anion to enter the middle of the cellulose sheets, with probable interference in the spacing of the cellulose microfibrils "tapes" or ribbons with the slight lifting of single cellulose fibers and eventual rupture of glycosidic bonds.

The acid concentration at which the process becomes effective has already been determined, in the analytical scheme of the nutritional value of food for ruminants—the van Soest method—in which the cellulose is depolymerized with sulfuric acid at a concentration of about 70% (w/v) at room temperature. The preparation of nanocellulose from microcrystalline cellulose (MCC) is generally carried out in the range of 56–64% sulfuric acid to promote the reduction of particle size. These results suggest that the combined action of several sulfate anions is necessary, reminiscent of the appearance of colligative properties, which can begin to "dope" the crystalline structure with multiple sulphate anions, leding to the disconnection of a single cellulose fiber, which in turn, make the glycosidic bond easier to hydrolyze to produce "loose ends" of fibers. The repetition of the process will have the final effect of de-crystallization with simultaneous partial hydrolysis of cellulose.

This mechanism was put to the test and fine-tuned in a large number of energy minimization tests in computational models performed with pairs of β-(1,4) octasaccharides with the most likely hydrogen bonding network, according to diffraction data of X-rays for cellulose allomorph I (see Figure 2a 1 to 3). A sequential model of approximating the sulfate anion and entering the crystalline network by replacing hydrogen bonds between a pair of monosaccharides in the chain with the sulfate oxygen's is depicted in Figure 2c, gives rise to a sequence, from which a summary is shown in Figure 2a). Based on total energy values, the entry of the sulphate anion (Figure 2, a-2) into the crystal structure is possible with apparent ease and small total energy change, and leads to a greater, but slight distancing of the pair of fibrils in the crystal sheet. To make this effect easier to see, a different diagram with equivalent steps is illustrated in Figure 2b), sequence 1-2-3. In this (a) they represent as simple cellulose fibrils, (b) with hydrogen bonds, (c) the sulfate anion entering the crystal bonding network. In Figure 2b-3, a single fibril begins to distance itself from the surface of the crystal, allowing its defibrillation and eventual hydrolysis.

In Figure 3, an octa-glucoside chain model is shown, with energy minimized, with and without the inclusion of sulfate anions in the crystalline cellulose network (A and B respectively). The distance measurement in both models results in a variation of the distance between the oxygen atom of the glycosidic bond in both situations of a value approximately 12% higher in the presence of anion sulfate. This result illustrates the easy intercalation of sulfate anion in the crystalline cellulose network.

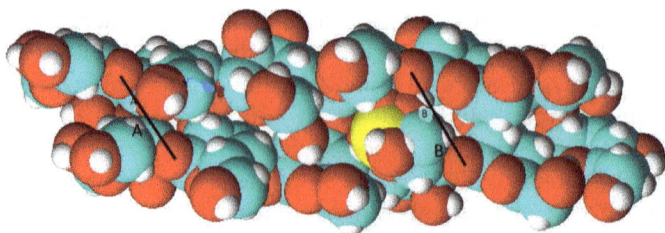

Figure 3. Cellulose crystal allomorph I with simplified hydrogen bonding network, optimized in the minimum of energy, without (A) and with an inclusion of sulfate anion (B). The dimensions of B are approximately 12% larger than A.

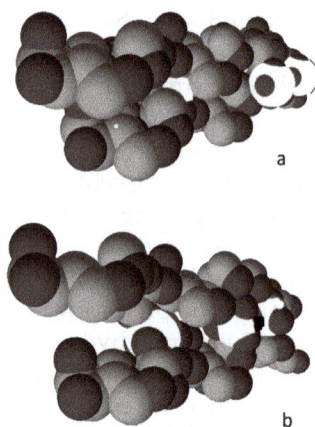

Figure 4. Microfibril exfoliation model (a-b) by joint action of multiple inclusion of sulfate anions (in white) in the hydrogen bonding network of crystalline cellulose.

In order to make microfibril exfoliation effective, the process becomes efficient only with an acid concentration greater than approximately 60–70% (w/v) which points to a similarity with the appearance of colligative properties. Thus, only the joint entry of several sulfate anions is operational in the physical separation of the microfibrils from the macrofibril array allowing for their rearrangement, hydrolysis or separation, as shown in Figure 4.

6. The methanol/sulfuric acid system

The results presented above seem to indicate that an effective process to decrystallize crystalline cellulose and promote its hydrolysis involves the use of concentrated sulfuric or phosphoric acid, or a mixture of both in conditions that allow the recovery of the acid. The crude biomass must be pre-treated for the removal of hemicellulose, since the hydrolyzed pentose rich sugars are not stable in the presence of strong acids.

Thus, two steps must be foreseen in the saccharification of cellulose: a first decrystallization, followed by the subsequent hydrolysis of the more loose fibril chain or amorphous cellulose. The use of cold concentrated sulfuric acid is particularly effective in decrystallization.

The use of concentrated sulfuric acid in biomass processing, described in the literature, uses almost exclusively the aqueous medium (Rivadeneira et al. 2019). The use of acids diluted in organic solvents, such as diethyl ether in impregnation methods, after removal of the solvent and subjected to mechanical treatments have shown to produce better results (Meine et al. 2012). Ethyl ether is a solvent not soluble in water, making it difficult to access predominantly hydrophilic polymers, especially hemicellulose and cellulose.

A detailed analysis of the solvents capable of being stable when mixed with concentrated sulfuric acid is shown in Figure 5. Ethanol would be a possible solvent. Alcohols will be one of the most desirable solvents to allow good contact between strong acids and biomass polymers. However, ethanol reacts with concentrated sulfuric acid (98%) at room temperature to produce ethylene gas. The process is an old, common method of preparing ethylene, a hormone that promotes fruit ripening. Ethylene is formed by the elimination of water catalyzed by concentrated sulfuric acid (Figure 5a) through the attack of the acid/sulfate anion on the hydrogen β of alcohol, discarding its possible use. In addition to solubility problems, diethyl ether also reacts with concentrated sulfuric acid, producing different decomposition products (Figure 5b). Both ethyl ether and ethanol have β-hydrogens, which allows their removal by catalytic elimination by sulfuric acid, giving rise to alkenes. Compounds unsaturated in acidic medium polymerize to give multiple products seen by the darkening of acid/ethanol solutions along with the evolution of gas, a method that can prepare some ethers. An important

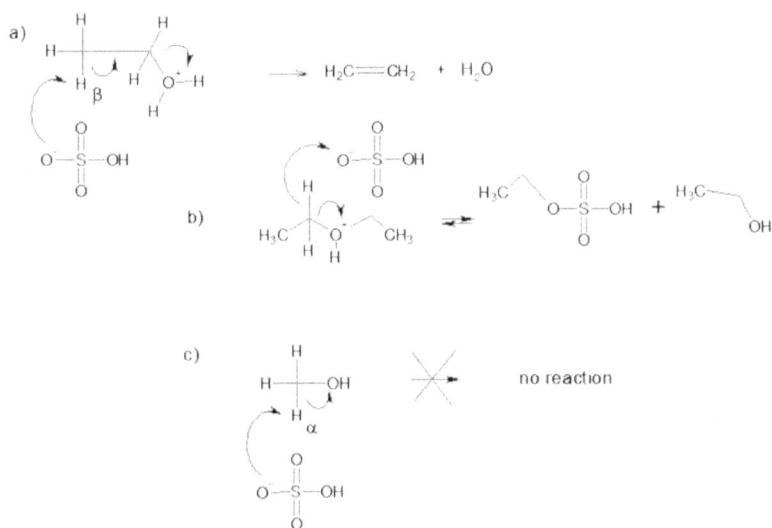

Figure 5. Reactivity of ethanol, methanol and ethyl ether with sulfuric acid. Methanol, without β hydrogens, is stable to the presence of concentrated solutions of sulfuric acid.

observation is the absence of β hydrogens in methanol, which must be stable in the presence of concentrated sulfuric acid (Figure 5c).

In fact, it is possible to mix concentrated sulfuric acid in all proportions with methanol without seeing any apparent reaction, in addition to the heat released by the exothermic dissolution. For the purposes mentioned in this work, solutions of concentrated sulfuric acid and methanol are stable up to 10–15°C, with no reaction product being observed. In some rare cases, a slight darkening occurs due to the presence of small contaminating compounds. As the dissolution is very exothermic, the solution must be prepared at low temperature, usually in ice-cooled methanol in a suitable laboratory environment and under care. The possible formation of methyl ether (CH3OCH3), which must be visualized with the release of a gas, has never been observed. In the present work, these results are published for the first time, together with results from some years of research using methanol/ concentrated sulfuric acid mixtures to promote the decrystallization of cellulose, but also its partial hydrolysis, a methodology that allows recovery of sulphuric acid and methanol, constituting an alternative to the use of ionic liquids. Figure 6 depicts the common processing to perform the decrystallization and hydrolysis of cellulose, exemplified with a sheet of filter paper. With minor modifications, the method can be used with any sample and type of biomass.

In Figure 6, in steps 1 and 2, the mixture of concentrated sulfuric acid (98%) in methanol is carried out with the solvent previously cooled in an ice bath 3. The acid/methanol ratio is, for example, 60/40 (v/v). The addition is slow to allow the heat to dissipate, stirring occasionally with a glass rod and reaching a temperature below 3°C. At this point, a rolled filter paper sheet (4) is immersed with the help of plastic tweezers and left in the solution for different periods of time, depending on the material. Using different types

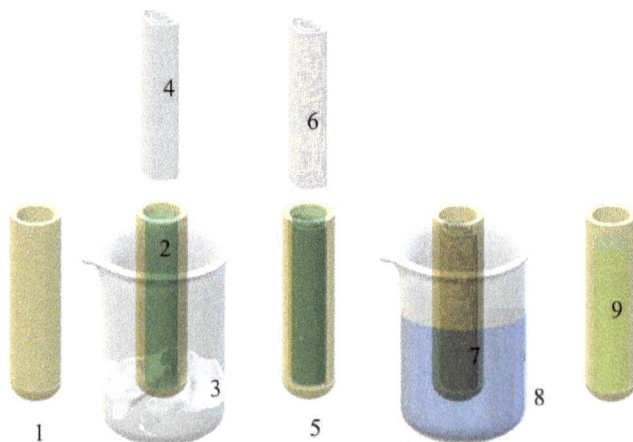

Figure 6. Exemplification of an experimental procedure to perform the decrystallization and hydrolysis of cellulose with a mixture of methanol/concentrated sulfuric acid, with the possibility of recovering sulfuric acid and methanol (see text for details).

of pre-treated biomass, different temperature/time settings can be used, the temperature range between –10 to 3°C are possible values with reaction times from 10 min to about one hour. After the designated period, the cellulose is decrystallized producing with filter paper a transparent gelatinous material (6), weakened in its mechanical resistance, but still possible to be removed by plastic tweezers or any other physical method at a certain point. Immediately, the gelatinous paper is placed in another tube containing pure methanol and washed repeatedly by mechanical movement (step 5 in Figure 6). Sometimes, a white precipitate is formed on the surface of the paper, its nature should be β(1,4) oligosaccharides insoluble in methanol. An example of such precipitates is identifiable in the photograph shown in Figure 7, in the form of a white precipitate on the surface of the paper.

The purpose of this step is to quickly remove most of the sulfuric acid from the paper or biomass sample. The next step is the hydrolysis of cellulose or biomass that is done under conditions similar conditions to those used to remove amorphous cellulose in the preparation of MCC (Shao et al. 2020) or other suitable conditions, steps 7 and 8 in Figure 6. A simple version is to use the remaining sulfuric acid absorbed in the paper, adding only water. The excess sulfuric acid dissolves in methanol and can be easily recovered, after a few batches, by simple distillation, both methanol and sulfuric acid being recovered in the process, which can be reused. To the decrystallized cellulose transferred in steps 7 and 8 of Figure 6, pure water or 2M hydrochloric acid is added, and it is boiled for a period of time from 5 to 30 minutes, depending on the type of biomass. In the case of paper, the hydrolytic treatment is continued until it is completely solubilized (step 9, see Figure 6). At this point, the sample is completely saccharified and able to follow the subsequent designed process.

Table 1 presents a compilation of results using this methodology, using different acid and temperature conditions, involving only the quantification

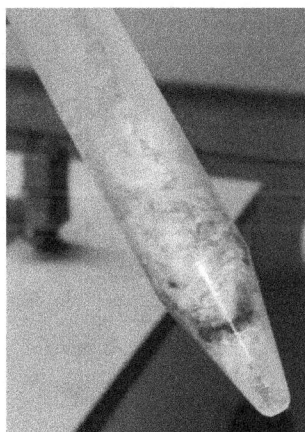

Figure 7. Formation of white solid precipitate after immersion of filter paper in pure methanol after de-crystallization with H2SO4/MeOH.

Table 1. Compilation of saccharification and filter paper hydrolysis results using the methanol/concentrated sulfuric acid system. The results of glucose and ethanol concentration are expressed as a percentage of the initial mass of paper.

Sample	Decrystalinization conditions			Saccharification conditions				Fermentation	Notes
	CH_3OH/H_2SO_4 (v/v)	Temp (°C)	Tim (min)	Temp (°C)	Tim (min)	Type/[ácido]	Sacchar. yield % Glucose	Conversion to ethanol (%)	
1 Starch	50/50	0	10	100	10	H_2SO_4/residual	~98	73.52 ± 3.33	Reference
2 Filter paper 1	50/50	16	30	100	10	H_2SO_4/residual	~9	5.31 ± 1.10	Acid adsorbed
3 Filter paper 1	50/50	16	45	100	10	H_2SO_4/residual	ND	4.40 ± 0.32	
4 Filter paper 1	50/50	0	5	100	10	H_2SO_4/residual	ND	5.12 ± 0.41	
5 Filter paper 1	50/50	0	15	100	10	H_2SO_4/residual	ND	6.13 ± 0.22	
6 Filter paper 1	50/50	0	30	100	10	H_2SO_4/residual	ND	4.33 ± 0.87	
7 Filter paper 1	40/60	16	30	100	10	H_2SO_4/residual	~12	5.33 ± 1.10	
8 Filter paper 1	40/60	16	5	100	10	H_2SO_4/residual	ND	4.12 ± 0.41	
9 Filter paper 1	40/60	16	10	100	10	H_2SO_4/residual	ND	4.56 ± 0.41	
10 Filter paper 1	40/60	16	20	100	10	H_2SO_4/residual	ND	4.14 ± 0.22	
11 Filter paper 1	40/60	0	5	100	10	H_2SO_4/residual	ND	7.62 ± 0.58	
12 Filter paper 1	40/60	0	10	100	10	H_2SO_4/residual	ND	3.97 ± 0.57	
13 Filter paper 1	40/60	0	20	100	15	H_2SO_4/residual	ND	2.75 ± 0.67	
14 Filter paper 1	40/60	0	10	100	15	H_2SO_4/residual	2.62 ± 0.21	ND	
15 Filter paper 2	40/60	0	10	100	15	2M HCl	7.22 ± 0.51	ND	
16 Filter paper 2	40/60	0	15	100	15	2M HCl	12.36 ± 0.87	ND	
17 Filter paper 2	40/60	0	20	100	15	2M HCl	91.55 ± 2.41	ND	
18 Filter paper 2	40/60	0	25	100	15	2M HCl	79.35 ± 2.76	ND	

ND. Not determined.

of simple sugars by the HPLC method, others including yeast/nutrient inoculation sequences and direct fermentation to obtain ethanol in order to verify the possible presence of any formation of fermentation inhibitors, namely furfural.

The results seem to indicate that with the use of filter paper, decrystallization at 0°C occurs relatively quickly. The thickness and porous nature of the filter paper facilitates the access of the acid. Hydrolytic tests using only residual sulfuric acid absorbed in the paper appear to be insufficient to promote complete hydrolysis of "amorphous" cellulose, taking into account the small yields obtained in ethanol after fermentation. However, obtaining ethanol means that in the decrystallization process little or no furfural is obtained, a result confirmed by analyzes in GC-MS. These, if formed, completely inhibit the fermentation of sugars, even in very low concentrations. The use of 2M hydrochloric acid in the hydrolysis stage constitutes a more efficient method of hydrolysis, and is suitable to obtain saccharification yields close to 90%. This methodology constitutes a viable procedure for saccharifying crystalline cellulose otherwise inaccessible, present extensively in many types of biomass.

7. References

Bakker, R. R., Gosselink, R. J. A., Maas, de Vrije T. and E. R. H. W. de Jong. 2004. Biofuel Production from Acid-Impregnated Willow and Switch grass. Agrotechnology and Food Innovations-Wageningen University and Research Centre. TechnoInvent BV, the Netherlands. 2nd World Conference on Biomass for Energy, Industry and Climate Protection, Rome, Italy.

Griffiths, B. S., Bracewell, J. M., Robertson, G. W. and D. E. Bignell. 2013. Pyrolysis-mass spectrometry confirms enrichment of lignin in the faeces of a wood-feeding termite, Zootermopsis nevadensis and depletion of peptides in a soil-feeder, Cubitermes ugandensis. Soil Biology & Biochemistry 57: 957–959.

Harmer, M. A., Fan, A., Liauw, A. and R. K. Kumar. 2009. A new route to high yield sugars from biomass: Phosphoric-sulfuric acid. Chem. Comm. 43: 6610–6612.

Meine, N., Rinaldi, R. and F. Schueth. 2012. Solvent-free catalytic depolymerization of cellulose to water-soluble oligosaccharides. Chemsuschem 5: 1449–1454.

Moe, Storker T., Janga, K. K., Hertzberg, T., Hagg, M. B., Oyaas, K. and N. Dyrset. 2012. Saccharification of lignocellulosic biomass for biofuel and biorefinery applications—A renaissance for the concentrated acid hydrolysis? Energy Procedia 20: 50–58.

Palaty, Z. and A. Zakova. 1996. Transport of sulfuric acid through anion-exchange membrane NEOSEPTA-AFN. Journal of Membrane Science 119: 183–190.

Reza, M. T., Rottler, E., Herklotz, L. and B. Wirth. 2015. Hydrothermal carbonization (HTC) of wheat straw: Influence of feed water pH prepared by acetic acid and potassium hydroxide. Bioresource Technology 182: 336–344.

Rivadeneira, J. P., Flavier, M. E. and F. R. P. Jr. Nayve. 2019. Optimization of acid and steam explosion pretreatment of cogon grass for improved cellulose enzymatic saccharification. Eurasian Chemico-Technological Journal 21: 143–147.

Schneider, L., Haverinen, J., Jaakkola, M. and U. Lassi. 2017. Pretreatment and fractionation of lignocellulosic barley straw by mechanocatalysis. Chemical Engineering Journal 327: 898–905.

Shao, X. Y., Wang, J., Liu, Z. T., Hu, N., Liu, M. and Y. W. Xu. 2020. Preparation and characterization of porous microcrystalline cellulose from corncob. Industrial Crops and Products 151: 112457.

Sun, Z. Y., Tang, Y. Q., Iwanaga, T., Sho, T. and K. Kida. 2011. Production of fuel ethanol from bamboo by concentrated sulfuric acid hydrolysis followed by continuous ethanol fermentation. Bioresource Technology 102: 10929–10935.

Wooley, R., Ma, Z. and N. H. L. Wang. 1998. A nine-zone simulating moving bed for the recovery of glucose and xylose from biomass hydrolysate. Industrial & Engineering Chemistry Research 37: 2699–3809.

Zhang, J. F., Wang, Y. X., Zhang, L. Y., Zhang, R. H., Liu, G. Q. and G. Cheng. 2014. Understanding changes in cellulose crystalline structure of lignocellulosic biomass during ionic liquid pretreatment by XRD. Bioresource Technology 151: 402–405.

Chapter 7

Microalgae Biomass as an Alternative to Fossil Carbons

Ferreira, Joana D, Martins, Clara B, Assunção, Mariana FG
and *Santos, Lilia MA*

1. Introduction

The constant interference of human activities in biogeochemical cycles and the increased use of fossil carbon are the main causes of global warming and the existing energy crisis (Milano et al. 2016, Raheem et al. 2018). Currently, 87% of the global CO_2 emitted by human activities results from the use of resources such as coal, oil and natural gas, contributing 43, 36 and 20% of sources, respectively. In the coming years, continued environmental deterioration is expected, as the human population is estimated to increase by up to 9 billion people by 2050 (Medipally et al. 2015, Raheem et al. 2018). Thus, it is essential to minimize carbon emissions using sustainable energy resources (Raheem et al. 2018). Biofuels have been used as an alternative to partly suppress dependence on fossil fuels (Velazquez-Lucio et al. 2018). Biofuels, which include biodiesel, bioethanol and biogas, are liquid or gaseous fuels derived from organic matter and constitute an alternative, green and renewable source for obtaining energy in a sustainable way (Velazquez-Lucio et al. 2018, Lakatos et al. 2019). Over the past decade, several types of biomass have gained interest as a new alternative source of biofuels and bioproducts (Enamala et al. 2018, Raheem et al. 2018). The production of biofuels has undergone evolution and has been classified essentially into four generations, depending on the source material and the process used (Alalwan et al. 2019, Lakatos et al. 2019). The first generation includes lipids and sugars from edible crops (for example, rapeseed, sugar

Coimbra Collection of Algae (ACOI), Department of Life Sciences, University of Coimbra, Portugal.

beet, corn or wheat), a source that generates great controversy for competing with human food needs. The second generation uses non-edible biomass (for example, lignocellulosic materials), however, there are limitations in the cost-benefit ratio involved in scale production at the commercial level (Raheem et al. 2018, Alalwan et al. 2019). The third generation focuses on the use of microorganisms such as microalgae, cyanobacteria, yeasts, fungi, and the fourth generation biofuels referes to the use of genetically modified microorganisms (Enamala et al. 2018, Alalwan et al. 2019). Microalgae are microscopic, single-celled or multicellular organisms, prokaryotic or photosynthetic eukaryotic with the ability to produce biomass and oxygen using sunlight as a source of energy and CO_2 as a carbon source (Patel et al. 2017). Its biomass has advantages for the production of biofuels in relation to the first and second generation that makes them an adequate alternative to agricultural crops and include: (1) capacity for cultivation in various lands such as arid and semi-arid , saline and other soils of low economic value; (2) less demand for nutrients in cultivation than those required by oilseeds; (3) possibility of using poor quality water supply, such as wastewater; (4) CO_2 requirements can be supplied from CO_2-rich flue gases, as they can tolerate NOx and SOx contained therein, and (5) a high photosynthetic efficiency and growth rate capable of allowing harvesting on a daily base (Demirbas and Demirbas 2010, Milano et al. 2016, Faried et al. 2017, Gouveia et al. 2017). However, despite the advantages of microalgae, their use in the production of biofuels has some disadvantages, namely the low concentration of biomass in solution and the low lipid content in some species. In addition, in some microalgae, the small size of their cells makes the collection and drying of biomass an expensive and energy-consuming process (Medipally et al. 2015). These limitations can be overcome with new techniques in development, such as the improvement of lipid harvesting, drying and extraction technology, the design of photobioreactors with greater photosynthetic efficiency or the development of biorefineries (Medipally et al. 2015, Chew et al. 2017).

This work presents a general review of the use of microalgae as a raw material for the production of biofuels, addressing the production of biodiesel and bioethanol from microalgae biomass, as well as the concept of microalgae biorefinery.

2. Energy production through the conversion of microalgal biomass

Microalgae biomass can be converted into various types of renewable biofuels, including biodiesel, bioethanol and biogas, among others. Three main production processes are available to produce biofuels from microalgae biomass: thermochemical, chemical and biochemical processes (Figure 1). The selection of the specific process depends on the desired biofuel. Thermochemical conversion includes pyrolysis, roasting, liquefaction and gasification. All of these technological processes use heat and catalysts to

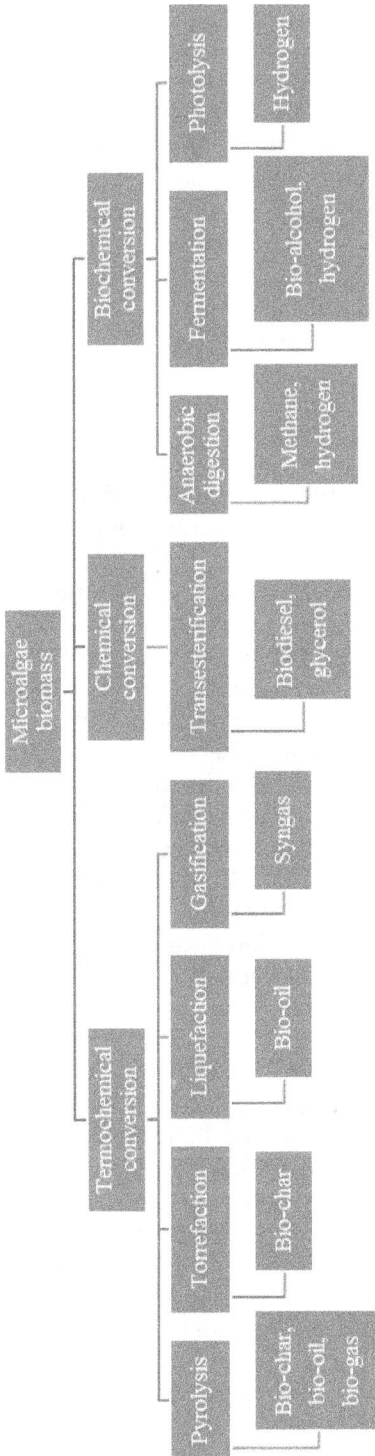

Figure 1. The processes for converting microalgal biomass into biofuels (adapted from Hossain 2019, Choo et al. 2020).

convert algae biomass into intermediate products, which are later converted into biofuels through chemical and biological routes (Raheem et al. 2018, Hossain 2019). The feasibility of producing biofuels from microalgae by thermochemical conversion may involve using syngas as an intermediate and obtaining products by converting it (Hossain 2019). The conversion of biomass may use processes such as transesterification, anaerobic digestion, fermentation and photolysis. These processes can convert microalgae biomass into biodiesel, methane, hydrogen and bioalcohols. Biochemical conversion processes are more environmentally friendly and consume less energy compared to other processes, however, at the moment, they are not suitable for large-scale production, as they have low conversion efficiency and are economically unfeasible (Choo et al. 2020).

2.1 Biodiesel from microalgae lipids

Biodiesel is an alternative fuel to fossil diesel and when mixed with it gives the diesel engine lower emissions of solid coal particles (Gouveia et al. 2017). If its physical properties are acquired according to the specification of the international standard it can be used in diesel engines up to concentrations up to 30% (Faried et al. 2017). Its total replacement may be made in some types of engine, but generally biodiesel is mixed with conventional diesel in different proportions, with a reduction in the use of fossil fuels (Demirbas and Demirbas 2010, Gouveia et al. 2017). The most important advantages that the use of biodiesel confer are the highest flash point, the biodegradability, the improved cetane number and the reduction of gaseous emissions harmful to the environment (Fazal et al. 2013). In addition, it is easy to obtain and available, renewable, non-flammable, non-toxic and ecological (Abbaszaadeh et al. 2012, Faried et al. 2017). Biodiesel consists of a mixture of fatty acid metil esters (FAME) obtained by transesterification of triacylglycerides (TAGs) and/or esterification of free fatty acids (FFAs) with alcohols, usually methanol, less often ethanol (Raheem et al. 2018). There are two methods of transesterification, the two-step method (conventional) and the method of direct transesterification. In the two-step method, lipid extraction from dry biomass is carried out by mechanical and/or chemical methods, before the transesterification and purification steps. The second method involves extracting and transesterifying lipid directly and simultaneously in dry biomass (Johnson and Wen 2009, Milano et al. 2016). In terms of agricultural production, biodiesel can be obtained from different types of crops, such as palm, soy, rapeseed and sunflower or from animal fat, being recognized as first generation biodiesel (Gouveia et al. 2017). However, first generation biodiesel generates controversy because of the competition of lipid consumption between the food and fuel chain, with the exception of the use of residual lipids.

Microalgae show advantages as an alternative to agricultural crops for the production of biodiesel. Associated with these advantages, biomass has a content in proteins, carbohydrates, carotenes that enriches its interest in use

Table 1. Lipid content of different microalgae (adapted from Demirbas and Demirbas 2011, D'Alessandro and Filho 2016, Sajjadi et al. 2018).

Microalgae	Oil content (% of dry biomass)
Ankistrodesmus gracilis	7.9–20.5
Auxenochlorella protothecoides	39.3
Botryococcus braunii	25.0–75.0
Chlorella sp.	28.0–32.0
Chlorella vulgaris	14.0–22.0
Crypthecodinium cohnii	20.0
Cylindriotheca sp.	16.0–37.0
Dunaliella primolecta	23.0
Euglena gracilis	14.0–20.0
Gymnodinium sp.	30.0
Isochrysis sp.	25.0–33.0
Monallanthus salina	> 20.0
Nannochloris sp.	20.0–35.0
Nannochloropsis sp.	31.0–68.0

as a raw material for obtaining other bioproducts. Microalgae can accumulate high values of lipid content, reaching in some cases up to 90% of their dry biomass (Table 1). Other microalgae species have different types of lipids and fatty acids combined up to values of about 40% of their biomass (Demirbas and Demirbas 2011).

Lipids are classified as polar or nonpolar according to their chemical structure and polarity. In microalgae, polar lipids, which include phospholipids and glycolipids, are present in cell membranes. Nonpolar or neutral lipids, used as an energy source, comprise acylglycerols (mono, di and tri) and free fatty acids (Sati et al. 2019). Triacylglycerols (TAGs) are the main target material for the production of biodiesel, especially if it has a lesser degree of unsaturation. Thus, microalgae with large amounts of lipids and relatively low amounts of TAGs are not considered suitable for the production of biodiesel (Sati et al. 2019). Some species of microalgae with lipids with a high TAG content, namely *Chlorella* sp., *Scenedesmus* sp. and *Nannochloropsis* sp., were studied for the production of biodiesel and subsequent applications of biofuels (Dong et al. 2016). The lipid biosynthesis in microalgae occurs in the chloroplast through the fatty acid pathway. (Gouveia et al. 2017). This is one of the main pathways of cellular metabolism, synthesizing fatty acids from acyl groups, necessary for the construction of the cell membrane and storage lipids. Apparently the process is similar to plant cells, the production of fatty acids requires acyl groups in the form of Acetyl-CoA, energy (ATP) and reducing agents such as NADPH, in a multi-enzymatic sequential process (Figure 2) (Khozin and Goldberg 2016). Many studies are focused

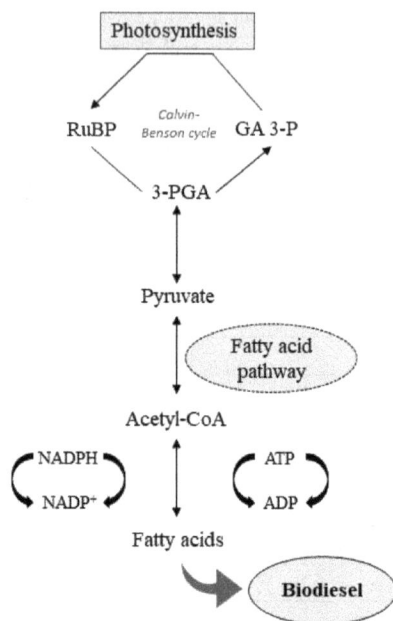

Figure 2. Fatty acid synthesis pathways in microalgae (adapted from Gouveia et al. 2017).

on identifying the lipid synthesis pathways in microalgae, namely on the identification and characterization of the main genes and key enzymes involved in the process (Yu et al. 2011, Gouveia et al. 2017).

The production of lipids, namely TAG's, by microalgae can be increased by combining several cultivation parameters, such as temperature, light intensity or nutrient limitation (Gouveia et al. 2017, Xiao et al. 2015). Under high light conditions, *Nannochloropsis ocenanica* IMET1 can produce large amounts of biomass and TAGs. On the other hand, *Ankistrodesmus falcatus* shows a high growth of biomass, but low content of lipids when subjected to high light intensity (George et al. 2014). The microalgae *Acutodesmus dismorphus* has a higher production of lipids and biomass at elevated temperatures, between 35 and 38°C, when compared to its growth at 25°C (Chokshi et al. 2015). Although microalgae have a high potential for the production of lipid, potentially convertible into biodiesel, it is necessary to overcome some challenges for commercialization and large-scale production to be viable. Currently, the costs associated with processes such as harvesting, drying and extracting biomass from microalgae are still high (Medipally et al. 2015). These challenges can be minimized with large-scale farming and harvesting systems with reduced cost per unit area. Examples are (1) obtaining more efficient photobioreactors with high biomass productivity at low cost (Kirrolia et al. 2013), (2) using nutrient-rich wastewater for biomass production (Gouveia et al. 2017) and, (3) the development of the biorefinery concept that could allow the joint use of lipids and co-products such as cell biomass (Milano et al. 2016).

2.2 Bioethanol from the fraction of microalgae carbohydrates

Bioethanol is the product of the fermentation of sugars such as glucose, fructose or sucrose by some microorganisms and has been recognized as being able to be used as a fuel, with characteristics of being sustainable, not contributing to CO_2 emissions, as it is non-toxic, biodegradable and produce a low content of pollutants for the environment (Jambo et al. 2016). As microalgae can accumulate up to 70% of their dry weight in carbohydrates under stress conditions (Table 2), it makes microalgae biomass a raw material, obtained by absorbing atmospheric CO_2, with potential for the production

Table 2. Carbohydrate production in microalgae (adapted from de Farias Silva et al. 2016, Phwan et al. 2018).

Microlagae	Carbohydrates (% DW)
Anabaena cylindrica	25.0–30.0
Aphanizomenon flos-aquae	23.0
Chlamydomonas reinhardtii	17.0
Chlorella sp.	19.5
Chlorella sorokiniana	35.7
Chlorella vulgaris	21.0
Chloroccum sp.	32.5
Dunaliella salina	32.0
Dunaliella tertiolecta	21.7
Euglena gracilis	14.0–18.0
Isochrysis zhangjiangensis	23.2–47.7
Isochrysis galbana	7.7–13.6
Isochrysis sp.	5.2–16.4
Nannochloropsis oceanica	22.7
Nannochloropsis oculata	8.0
Pavlova lutheri	28.2
Porphyridium cruentum	40.0
Prymnesium parvum	25.0–33.0
Scenedesmus dimorphus	21.0–52.0
Scenedesmus obliquus	10.0–17.0
Spirulina platensis	31.2
Spirogyra sp.	33.0–64.0
Spirulina sp.	20.0
Tetraselmis maculata	15.0
Tetraselmis suecica	15.0–50.0
Tetraselmis sp.	24.0

Table 3. Comparison of ethanol production between microalgae and crops (adapted from Gouveia et al. 2011).

Source	Ethanol yield (L/ha)
Corn stover	1.0–1.4
Wheat	2.6
Cassava	3.3
Sweet sorghum	3.0–4.1
Corn	3.5–4.0
Sugar beet	5.0–6.7
Sugar cane	6.2–7.5
Switch grass	10.8
Microalgae	46.8–140.3

of bioalcohols such as ethanol or butanol, with an advantage in comparison with conventional crops, as they present higher productivity, see Table 3 (Gouveia et al. 2011, Smachetti et al. 2018).

Its carbohydrate content occurs mainly in the form of starch and low crystallinity cellulose, which are easily hydrolyzed to fermentable sugars through microbial action. The microalgae cell wall composition contributes to this, which is different from the plants used in lignocellulosic cultures, due to its low content or absence of lignin and hemicellulose, making it an interesting biotechnological raw material (Phwan et al. 2018, Smachetti et al. 2018). The production of bioethanol is carried out in three stages: pretreatment, saccharification and fermentation (Velazquez-Lucio et al. 2018). In recent years, several processes involved in the production of bioethanol such as the selection of microalgae strains, optimization of pre-treatment methods to efficiently break down their cell wall, selection of the fermentative microorganism and cost of staggering processes, have been the subject of intense research focused on increasing bioethanol production with the lowest economic and environmental costs (Phwan et al. 2018). Currently, there are two methods for breaking down carbohydrate polymers into simple sugars or monomers and converting them into bioethanol, based on research carried out in an attempt to convert biomass into hydrolysis and separate fermentation (SHF) and simultaneous saccharification and formation (SSF) (Smachetti et al. 2018). These methods manage to obtain better yields in conversion to ethanol, in an economically viable way (Jambo et al. 2016). In SHF, hydrolysis to break down carbohydrates into monomeric sugars is carried out first, followed by the fermentation of the chosen sugars. Instead, in the SSF, the hydrolysis and fermentation processes are carried out simultaneously, combining yeast and enzymes, allowing a quick conversion of sugars into bioethanol. The SSF seems to have a greater consensus due to the fact that it is a simultaneous, simpler process, the cost of producing bioethanol is lower and the production rate is higher (Alfani et al. 2000, Dahnum et al. 2015). Studies with several strains of microalgae for the production of bioethanol show promising results especially in biomass of microalgae rich in carbohydrates (Silva et al. 2016). The manipulation of cultivation conditions, such as light intensity, temperature and nutrient limitation, is one of the strategies to induce greater productivity of carbohydrates in microalgae (Markou et al. 2012, Shuba et al. 2018). Some examples are the studies carried out in *Scenedesmus* and *Chlorella*, to enhance the production of carbohydrates and, consequently, increase the yield of bioethanol production (Rizza et al. 2017). *Chlorella vulgaris* FSP-E and ESP-6 grown under nitrogen depletion conditions exhibited an increase in carbohydrates from 15–20% to 49–54% (Ho et al. 2013). *C. vulgaris* CCALA 924, growing under suitable conditions of phosphorus and nitrogen, but sulfur depletion, shown that this limitation induces the accumulation of starch in the biomass, up to values close to 60% of dry weight (DW), 50% higher than that without sulfur deprivation (Branyikova et al. 2011). *C. vulgaris* P12 was grown with a limit concentration of Fe (III) (FeNa-EDTA) and/or urea, and it was found that only the depletion of the nitrogen source (urea) increases the starch content up to about 40% (DW) cell biomass (Dragone et al. 2011). In *Tetraselmis subcordiformis* the relationship between nitrogen (0–11 mM nitrate) and sulfur (0–0.8 mM sulfate) limitations was studied, showing that nitrogen limitation has a more significant role in starch accumulation than limitation sulfur or even than the combination of both (Yao et al. 2012). Studies on the potential for bioethanol production from microalgae over the years increased and various technologies developed with the aim of making these processes more sustainable (Jambo et al. 2016, de Farias Silva et al. 2016).

3. Microalgae in biorefinery

The term biorefinery has been present in the scientific literature since 2001 (Wyman 2001), several definitions for this term have been proposed. According to the International Energy Agency (IAE), bio-refining is the processing of biomass in a sustainable manner with the aim of obtaining marketable products and energy (Sonnenberg et al. 2007). One of the objectives of the biorefinery is to mitigate the emission of greenhouse gases, since fossil fuels have contributed strongly to global warming (Chew et al. 2017). Thus, one of the goals of biorefinery based on microalgae is, at the same time, to remove CO_2 from the atmosphere, to develop sustainable production methods to obtain biofuels, bioenergy and bio-products with high added value through isolation and/or transformation of components of its biomass (Chew et al. 2017, Raheem et al. 2018). The main stages of biorefineries are divided into two processes, upstream processing (USP) and downstream processing (DSP). The UPS includes all the initial steps. Four essential factors contribute to USP's efficiency, the strain of the microorganism, the supply of carbon dioxide, nutrients and the source of light. In the case of microalgae biorefineries, UPS factors have been studied and are better known over time

(Vanthoor-Koopmans et al. 2013, Aravantinou and Manariotis 2016, Chew et al. 2017). The next steps are included in the DSP which involves biomass/ product separation, product purification, concentration and, finally, its sale (Lin and Luque 2014). Examples of DSP in microalgae are the methods of extraction and purification to obtain marketable substances (Chew et al. 2017). In the concept of microalgae biorefinery, it should be noted that conventional techniques, such as lipid extraction, have high production costs and, therefore, are economically unfeasible, making it necessary to develop and integrate several steps that improve processes in terms of economy, simplicity and ease of processing (Chew et al. 2017). Other examples, such as pre-treatment processes that consume less energy, harvesting and dewatering with lower associated costs, need to be developed, since these two processes reach up to 50% of the total production cost of biomass (Lee et al. 2015, Raheem et al. 2018). In addition, a low-cost and low-energy extraction method is required (Lee et al. 2015). In recent years, the variety of bioproducts obtained from microalgae biomass has been a focus of interest, and the concept of microalgae biorefineries is currently a very attractive area of research (Lim et al. 2012, Raheem et al. 2018). Not only are lipids extracted in the microalgae biorefinery, other metabolites, including pigments, carbohydrates, proteins, enzymes or minerals can also be extracted for development in refined products for various applications (Figure 3) (Yen et al. 2013, Dickinson et al. 2017, Raheem et al. 2018). For example, in addition to the use of lipids for energy, some long-chain fatty acids found in microalgae are important dietary supplements. In addition, several proteins and pigments have been competent in the pharmaceutical industry for the treatment of diseases and human and animal nutrition. In addition, microalgal carbohydrates can be used as a carbon source to replace the traditional carbohydrates from the fermentation industry (Yen et al. 2013).

Along with the lipid that can be converted into biodiesel, other microalgae compounds should be used to improve the economy of the biorefinery process. The fully integrated production of biodiesel, bioethanol and carotenes can be a key solution for a biorefinery without residues from biomass from microalgae. In the near future, the microalgae biorefinery concept for the production of bioproducts and biofuels is expected to play an important role in the bio-based economy (Raheem et al. 2018).

4. Final considerations

Microalgae as a raw material for the production of biofuels has been the focus of intense research in order to replace them as first and second generation raw materials, namely oil crops. As described, microalgae biomass can be converted into various types of renewable bioproducts and biofuels. Biodiesel and bioethanol are the biofuels obtained from microalgae with more research underway/carried out, biodiesel for the high content of lipids and bioethanol for the high content of carbohydrates that are easy to saccharify. However, for

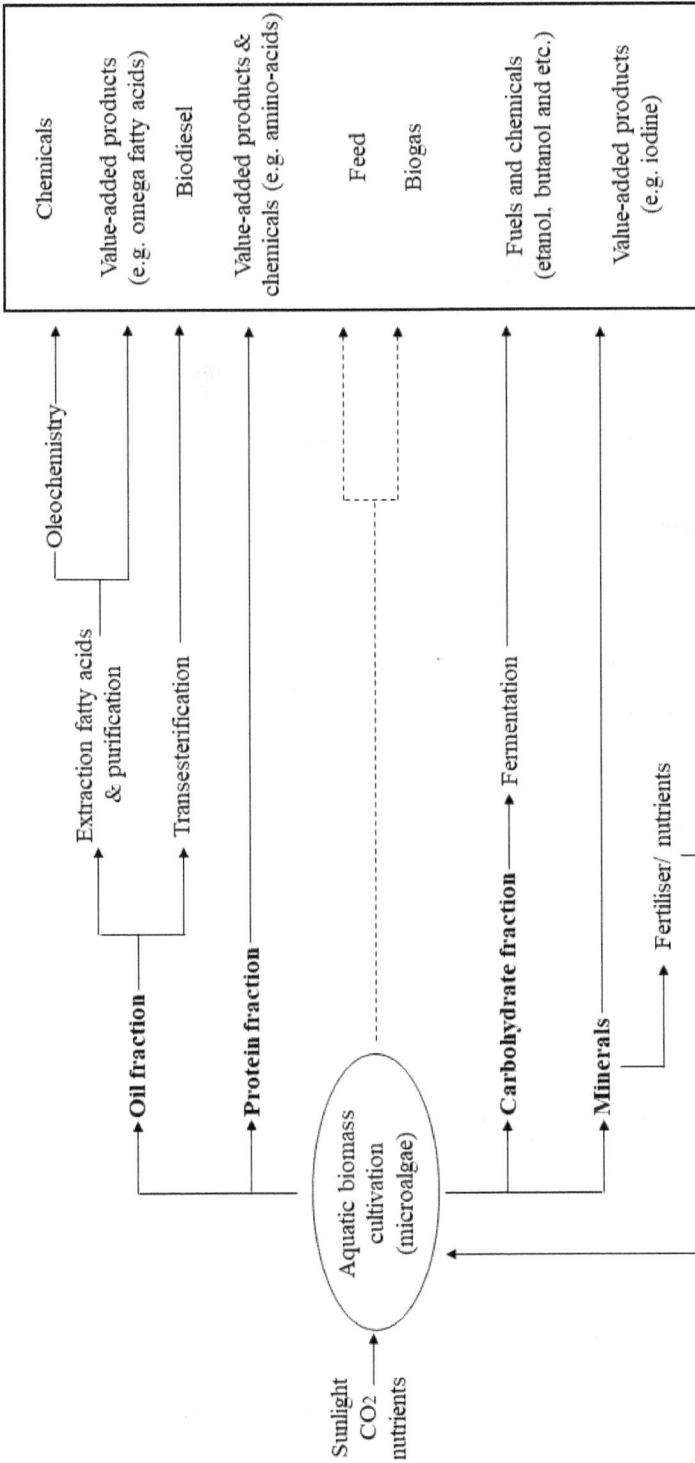

Figure 3. The concept of microalgae biorefinery (adapted from Chew et al. 2017).

both biofuels to be commercially viable, several challenges need to be faced and resolved. One solution involves the concept of microalgae biorefinery, which allows profitable use of microalgae biomass for biofuels concurrently with the exploitation of other co-products to improve the economic process.

5. References

Abbaszaadeh, A., Ghobadian, B., Omidkhah, M. R. and G. Najafi. 2012. Current biodiesel production technologies: A comparative review. Energy Conversion and Management 63: 138–148.

Alalwan, H. A., Alminshid, A. H. and H. A. Aljaafar. 2019. Promising evolution of biofuel generations. Subject review. Renewable Energy Focus 28: 127–139.

Alfani, F., Gallifuoco, A., Saporosi, A., Spera, A. and M. Cantarella. 2000. Comparison of SHF and SSF processes for the bioconversion of steam-exploded wheat straw. Journal of Industrial Microbiology and Biotechnology 25: 184–192.

Aravantinou, A. F. and I. D. Manariotis. 2016. Effect of operating conditions on *Chlorococcum* sp. growth and lipid production. Journal of Environmental Chemical Engineering 4(1): 1217–1223.

Brányiková, I., Maršálková, B., Doucha, J., Brányik, T., Bišová, K., Zachleder, V. and M. Vítová. 2011. Microalgae novel highly efficient starch producers. Biotechnology and Bioengineering 108: 766–776.

Chew, K. W., Yap, J. Y., Show, P. L., Suan, N. H., Juan, J. C., Ling, T. C. and J. S. Chang. 2017. Microalgae biorefinery: high value products perspectives. Bioresource Technology 229: 53–62.

Chokshi, K., Pancha, I., Trivedi, K., George, B., Maurya, R., Ghosh, A. and S. Mishra. 2015. Biofuel potential of the newly isolated microalgae *Acutodesmus dimorphus* under temperature induced oxidative stress conditions. Bioresource Technology 180: 162–171.

Choo, M. Y., Oi, L. E., Ling, T. C., Ng, E. P., Lee, H. V. and J. C. Juan. 2020. Conversion of microalgae biomass to biofuels. *In*: Microalgae Cultivation for Biofuels Production. Academic Press, pp. 149–161.

D'Alessandro, E. B. and N. R. Antoniosi Filho. 2016. Concepts and studies on lipid and pigments of microalgae: A review. Renewable and Sustainable Energy Reviews 58: 832–841.

Dahnum, D., Tasum, S. O., Triwahyuni, E., Nurdin, M. and H. Abimanyu. 2015. Comparison of SHF and SSF processes using enzyme and dry yeast for optimization of bioethanol production from empty fruit bunch. Energy Procedia 68: 107–116.

de Farias Silva, C. E. and A. Bertucco. 2016. Bioethanol from microalgae and cyanobacteria: A review and technological outlook. Process Biochemistry 51: 1833–1842.

Demirbas, A. and M. F. Demirbas. 2010. Algae Energy: Algae as a New Source of Biodiesel. Springer Science & Business Media.

Demirbas, A. and M. F. Demirbas. 2011. Importance of algae oil as a source of biodiesel. Energy Conversion and Management 52: 163–170.

Dickinson, S., Mientus, M., Frey, D., Amini-Hajibashi, A., Ozturk, S., Shaikh, F. and M. M. El-Halwagi. 2017. A review of biodiesel production from microalgae. Clean Technologies and Environmental Policy 19: 637–668.

Dong, T., Knoshaug, E. P., Pienkos, P. T. and L. M. Laurens. 2016. Lipid recovery from wet oleaginous microbial biomass for biofuel production: a critical review. Applied Energy 177: 879–895.

Dragone, G., Fernandes, B. D., Abreu, A. P., Vicente, A. A. and J. A. Teixeira. 2011. Nutrient limitation as a strategy for increasing starch accumulation in microalgae. Applied Energy 88(10): 3331–3335.

Enamala, M. K., Enamala, S., Chavali, M., Donepudi, J., Yadavalli, R., Kolapalli, B. and C. Kuppam. 2018. Production of biofuels from microalgae—A review on cultivation, harvesting, lipid extraction, and numerous applications of microalgae. Renewable and Sustainable Energy Reviews 94: 49–68.

Faried, M., Samer, M., Abdelsalam, E., Yousef, R. S., Attia, Y. A. and A. S. Ali. 2017. Biodiesel production from microalgae: Processes, technologies and recent advancements. Renewable and Sustainable Energy Reviews 79: 893–913.

Fazal, M. A., Haseeb, A. S. M. A. and H. H. Masjuki. 2013. Investigation of friction and wear characteristics of palm biodiesel. Energy Conversion and Management 67: 251–256.

George, B., Pancha, I., Desai, C., Chokshi, K., Paliwal, C., Ghosh, T. and S. Mishra. 2014. Effects of different media composition, light intensity and photoperiod on morphology and physiology of freshwater microalgae Ankistrodesmus falcatus–A potential strain for bio-fuel production. Bioresource Technology 171: 367–374.

Gouveia, L. 2011. Microalgae as a feedstock for biofuels. pp. 1–69. *In*: Microalgae as a Feedstock for Biofuels. Springer, Berlin, Heidelberg.

Gouveia, L., Oliveira, A. C., Congestri, R., Bruno, L., Soares, A. T., Menezes, R. S. and I. Tzovenis. 2017. Biodiesel from microalgae. Microalgae-based biofuels and bioproducts. Woodhead Publishing, 235–258.

Ho, S. H., Huang, S. W., Chen, C. Y., Hasunuma, T., Kondo, A. and J. S. Chang. 2013. Bioethanol production using carbohydrate-rich microalgae biomass as feedstock. Bioresource Technology 135: 191–198.

Hossain, S. Z. 2019. Biochemical conversion of microalgae biomass into biofuel. Chemical Engineering & Technology 42(12): 2594–2607.

Jambo, S. A., Abdulla, R., Azhar, S. H. M., Marbawi, H., Gansau, J. A. and P. Ravindra. 2016. A review on third generation bioethanol feedstock. Renewable and Sustainable Energy Reviews 65: 756–769.

Johnson, M. B. and Z. Wen. 2009. Production of biodiesel fuel from the microalga Schizochytrium limacinum by direct transesterification of algal biomass. Energy Fuels 23: 5179–83.

Khozin-Goldberg, I. 2016. Lipid metabolism in microalgae. *In*: The Physiology of Microalgae. Springer, Cham, pp. 431–484.

Kirrolia, A., Bishnoi, N. R. and R. Singh. 2013. Microalgae as a boon for sustainable energy production and its future research & development aspects. Renewable and Sustainable Energy Reviews 20: 642–656.

Lakatos, G. E., Ranglová, K., Manoel, J. C., Grivalský, T., Kopecký, J. and J. Masojídek. 2019. Bioethanol production from microalgae polysaccharides. Folia Microbiologica, 1–18.

Lee, O. K., Seong, D. H., Lee, C. G. and E. Y. Lee. 2015. Sustainable production of liquid biofuels from renewable microalgae biomass. Journal of Industrial and Engineering Chemistry 29: 24–31.

Lin, C. and R. Luque. (eds.). 2014. Renewable resources for biorefineries (No. 27). Royal Society of Chemistry.

Markou, G., Angelidaki, I. and D. Georgakakis. 2012. Microalgal carbohydrates: An overview of the factors influencing carbohydrates production, and of main bioconversion technologies for production of biofuels. Applied Microbiology and Biotechnology 96: 631–645.

Medipally, S. R., Yusoff, F. M., Banerjee, S. and M. Shariff. 2015. Microalgae as sustainable renewable energy feedstock for biofuel production. BioMed Research International, 2015.

Milano, J., Ong, H. C., Masjuki, H. H., Chong, W. T., Lam, M. K., Loh, P. K. and V. Vellayan. 2016. Microalgae biofuels as an alternative to fossil fuel for power generation. Renewable and Sustainable Energy Reviews 58: 180–197.

Patel, A., Gami, B., Patel, P. and B. Patel. 2017. Microalgae: Antiquity to era of integrated technology. Renewable and Sustainable Energy Reviews 71: 535–547.

Phwan, C. K., Ong, H. C., Chen, W. H., Ling, T. C., Ng, E. P. and P. L. Show. 2018. Overview: comparison of pretreatment technologies and fermentation processes of bioethanol from microalgae. Energy Conversion and Management 173: 81–94.

Raheem, A., Prinsen, P., Vuppaladadiyam, A. K., Zhao, M. and R. Luque. 2018. A review on sustainable microalgae based biofuel and bioenergy production: Recent developments. Journal of Cleaner Production 181: 42–59.

Rizza, L. S., Smachetti, M. E. S., Do Nascimento, M., Salerno, G. L. and L. Curatti. 2017. Bioprospecting for native microalgae as an alternative source of sugars for the production of bioethanol. Algal Research 22: 140–147.

Sajjadi, B., Chen, W. Y., Raman, A. A. A. and S. Ibrahim. 2018. Microalgae lipid and biomass for biofuel production: A comprehensive review on lipid enhancement strategies and their effects on fatty acid composition. Renewable and Sustainable Energy Reviews 97: 200–232.

Sati, H., Mitra, M., Mishra, S. and P. Baredar. 2019. Microalgal lipid extraction strategies for biodiesel production: A review. Algal Research 38: 101413.

Shuba, E. S. and D. Kifle. 2018. Microalgae to biofuels: "Promising" alternative and renewable energy, review. Renewable and Sustainable Energy Reviews 81: 743–755.

Smachetti, M. E. S., Rizza, L. S., Coronel, C. D., Do Nascimento, M. and L. Curatti. 2018. Microalgal biomass as an alternative source of sugars for the production of bioethanol. Principles and Applications of Fermentation Technology, 351.

Sonnenberg, A., Baars, J. and P. Hendrickx. 2007. IEA Bioenergy Task 42 Biorefinery.

Vanthoor-Koopmans, M., Wijffels, R. H., Barbosa, M. J. and M. H. M. Eppink. 2013. Biorefinery of microalgae for food and fuel. Bioresource Technology 135: 142–149.

Velazquez-Lucio, J., Rodríguez-Jasso, R. M., Colla, L. M., Sáenz-Galindo, A., Cervantes-Cisneros, D. E., Aguilar, C. N. and H. A. Ruiz. 2018. Microalgal biomass pretreatment for bioethanol production: A review. Biofuel Research Journal 5: 780–791.

Wyman, C. E. 2001. Economics of a biorefinery for coproduction of succinic acid, ethanol, and electricity. Abstracts of Papers of the American Chemical Society. 221 P. U119–U119. Part 1 Meeting Abstract: 72-BIOT.

Xiao, Y., Zhang, J., Cui, J., Yao, X., Sun, Z., Feng, Y. and Q. Cui. 2015. Simultaneous accumulation of neutral lipids and biomass in Nannochloropsis oceanica IMET1 under high light intensity and nitrogen replete conditions. Algal Research 11: 55–62.

Yao, C., Ai, J., Cao, X., Xue, S. and W. Zhang. 2012. Enhancing starch production of a marine green microalga Tetraselmis subcordiformis through nutrient limitation. Bioresource Technology 118: 438–444.

Yen, H. W., Hu, I. C., Chen, C. Y., Ho, S. H., Lee, D. J. and J. S. Chang. 2013. Microalgae-based biorefinery–from biofuels to natural products. Bioresource Technology 135: 166–174.

Yu, W. L., Ansari, W., Schoepp, N. G., Hannon, M. J., Mayfield, S. P. and M. D. Burkart. 2011. Modifications of the metabolic pathways of lipid and triacylglycerol production in microalgae. Microbial Cell Factories 10: 91.

Chapter **8**

The Phase Inversion in the Preparation of Batch Biodiesel from Triglycerides and Methanol

1. Introduction

Humanity's dependence on fossil fuels as the main source of energy and raw materials has caused major environmental climate changes. Reducing this dependency is an urgent goal to be achieved to mitigate climate change. The finding of important new oil reserves in the subsoil is increasingly scarce (Miller and Sorrel 2014), which gives us an idea of the huge amount already explored and used by humanity. The electric replacement of automobiles has begun to gain momentum and should be completed by the middle of the century; however, the demand for fossil fuels will continue for a long time to come. Diesel is a type of fuel and engine that is very widespread, especially in the transport of heavy cargo. Its burning emits particles that are harmful to public health at the level of the respiratory system. One way to reduce these harmful emissions and comply with the levels of environmental legislation is to add biodiesel to diesel (Canakci et al. 2006). The use of biodiesel itself is ecologically correct because of its neutral contribution to CO_2 emissions (Hill et al. 2006), especially if its production originates from residual lipids such as from the human food chain, namely frying, or from renewable sources, such as algae biomass that do not compete for land use (Assunção et al. 2017, Varejão and Nazaré 2017, Martins et al. 2016, Daroch et al. 2013). In particular, converting residual edible oils into biodiesel is an appropriate way to find a useful destination for waste that is otherwise problematic. In this chapter, results obtained in a pilot unit for preparing biodiesel by transesterification of residual oils from student canteens, in batch reaction cycles, under conditions of homogeneous catalysis with KOH are reported (Vicente 2004, Leung 2010, Meher 2006).

The experimental conditions involved reaction times 2–12 h, temperatures from 20 to 60°C, a molar ratio of oil to methanol of 1: 6, and the presence of the KOH catalyst in concentrations of 0.5 to 1.0% (w/v). These conditions proved to result in conversion yields greater than 97% of FAME from the triacylglycerols in a reaction time of about 1 h at 60°C. Most of the results of biodiesel preparation in the literature refer to reactions made on a laboratory scale with pure oils, in real production with used oils new problems arise, such as the presence of contaminants in the lipid, namely water and/or free fatty acids (Vicente 2004, Leung 2010, Meher 2006, Talebian-Kiakalaieh et al. 2013). The desirable temperature of the reactor is 60°C, however heating to this temperature can consume considerable amounts of energy, especially in winter. This led the use of lower temperatures in a large number of batches, and it was found that the reaction proceeds with an acceptable speed at temperatures above 25°C, but with a longer reaction time, of about from 12–16 h. Other variables that were found to affect the yield and the total conversion time of oils used in FAME preparation was the different origin of methanol and especially its water content and the addition of recovered methanol.

The reaction medium at the beginning of the process consists of two phases, the oil as the denser phase and the methanol/KOH solution as the upper phase. With strong initial agitation promoted by vertical blade impeller (Csernicaa and Hsu 2012), the medium forms an emulsion which facilitates the beginning of the reaction. In a more advanced stage of the reaction, a new two-phase system is formed, now with the densest phase whose composition is a mixture of glycerol/methanol (sometimes described here as a glycerol rich phase) and the upper phase whose components are mainly FAME and oil that has not yet reacted. Despite the large number of published works on the topic of biodiesel preparation, few or no results are found in the literature on the issue of the formation of the glycerol/methanol phase and its characterization in terms of the concentration of all its components. This knowledge gap is even more surprising since the final formation of these phases makes it difficult to obtain yields close to those corresponding to the complete reaction (Csernica and Hsu 2013b). This is because the catalytic species, sodium methoxide, has a greater affinity for the glycerol/methanol phase avoiding contact with the unconverted oil in the upper phase, which is a phenomenon that hinders the phase transfer required for the reaction. To achieve conversion values in biodiesel required by the standard norm (% FAME > 96.5), additional finishing steps, such as catalyst addition or temperature increase are required. From results obtained with the completion of a high number of batches of biodiesel preparation, data on the behavior of the phases involved in the reaction coordinate associated with the FAME conversion degree are available. From these data, it was possible to build a mathematical model to predict the composition of the system when phase inversion occurs together with its dependence on the initial concentration of the reagents. Detailed analysis of the model results makes it evident that the formation of the glycerol/methanol phase does not mean the completion of the reaction. The measurement of the density of this phase proved to be a simple and convenient way of evaluating the degree of conversion of oil into FAME, by a simple technique accessible to all—the measurement of the lower phase density of glycerol/methanol.

2. Details of the pilot unit and operating conditions

A 400 L stainless steel cylindrical reactor with a water jacket was used to allow heating and equipped with a vertical paddle stirrer and thermometer. The paddle stirrer was connected to a programmable timer. The other components of the plant were a water washing tank (400 L), a distillation unit for the separation and recovery of methanol excess in the glycerol/methanol phase, liquid pumps, and mechanical stirrers for methoxide solution preparation.

The evolution of the reaction was followed by HPLC analysis, under the conditions described below.

The residual edible oils were collected in the canteens on the campus of the Escola Superior Agrária de Coimbra (ESAC/IPC) and subjected to gravitational sedimentation of the solids and water present for at least a period of 7 days. The oil was used only if it had a clear and transparent appearance. The acidity and fatty acid composition (fatty acid composition profile similar to soybean oil) were analyzed in each batch. In a typical test, 300 L of clarified residual cooking oil, with a titrimetric acidity to NaOH less than 1.0%, were pumped into the 400 L reactor. In a plastic pot with a stopper (100 L), potassium hydroxide pellets (200 g) were dissolved in 75 L of methanol (water content less than 100 ppm), with stirring by rotation to dissipate the heat of dissolution. The triglyceride/methanol molar ratio used was approximately 1/6. After complete dissolution, potassium methoxide in methanol was added to the reactor, which is then closed. Strong mixing of the mixture was carried out for ten minutes, with frequent inversion of the rotation of the blades to form an emulsion. The reaction was usually left for periods of 14 h or overnight at room temperature, if above 25°C; with lower temperatures, heating through a circulating water jacket was switched on to maintain a temperature of at least 25°C. The reaction medium was vigorously stirred hourly (10 min) by means of a programmable timer. The FAME yield was measured at the end of the reaction period by HPLC and if greater than 96.0%, the lower glycerol/methanol phase was transferred, via manual valves, to the distillation unit to recover excess methanol. The upper phase was transferred to a 400 L vessel and washed with a shower of water under pressure, until the washes, after resting, showed a transparent appearance. The aqueous phase was discarded. After 3–4 h of rest, the upper phase of clarified biodiesel was transferred to a storage tank. An intermediate emulsion phase forms in the washing process and is transferred to a separate vessel, in which phase separation occurs after longer periods of time, such as a week.

3. Preparation of biodiesel from residual food oils

3.1 *Analytical techniques in monitoring the transesterification reaction*

The residual edible oils supplied by the canteens have very low acid values, in the range of 0.3–1.0%, being converted into the fatty acid methyl ester (FAME) by means of the transesterification reaction in a single step, with catalysis with potassium methoxide. The availability of analytical methods to assess the degree of conversion of triglycerides to FAME is essential to determine the end of the reaction, in order to respect the parameters defined in the standard for biodiesel. The method used was that of determining triolein in olive oil (EU 1991), an HPLC method. This was previously verified as adequate to accompany the transesterification reaction with authentic samples of edible oils, biodiesel and mixtures of both. The method can distinguish the initial mixture of triglycerides from the end products of the reaction, the methyl esters of fatty acids (FAME). Both free fatty acids and mono- and diglycerides have chromatographic bands that elute at distant retention times from the bands corresponding to FAME, their occurrence being transient. In parallel, regular analyzes of the final product were carried out using the EN14105 method, using a new gas chromatography (GC) injection technique (Varejão and Cruz Costa 2015) in order to quantify the mono- and diglycerides content. Figure 1 shows a

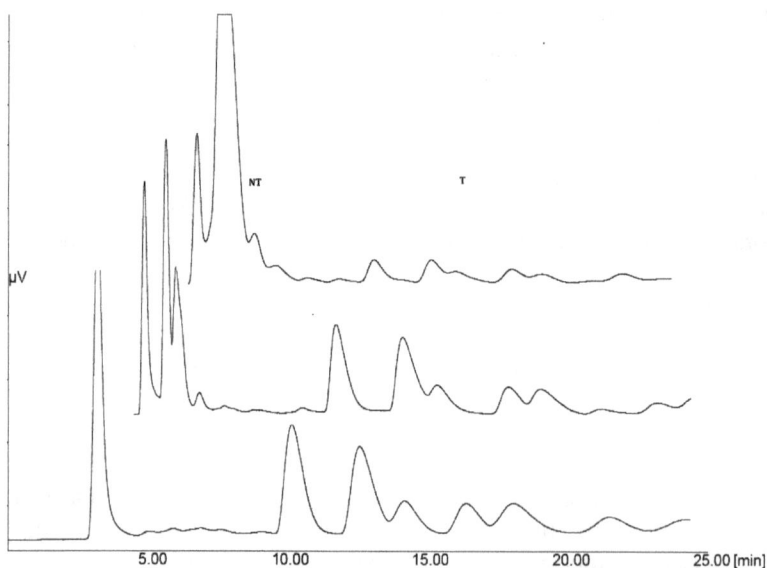

Figure 1. Sequence of HPLC chromatograms in the conversion of used edible oil in the corresponding fatty acids methyl esters. Below is the chromatogram of the residual edible oil, in the middle a sample from a batch with partial conversion to FAME and above a sample in an advanced stage of the reaction. Triglycerides (T) are bands with retention time (tr) between 10–26 minutes, and non-triglycerides (NT—FAME added to free fatty acids) the bands with retention time between 4–8 minutes. The first peak in the chromatograms is the system peak.

sequence of HPLC chromatograms in the conversion of a sample of used cooking oil to FAME under the conditions mentioned above.

3.2 Yield in FAME and formation of the glycerol/methanol phase

In the batch conversion of residual oil to biodiesel, at the beginning, the reaction medium shows two phases, the upper one with methanol and the lower one the oil. Methanol is generally referred to as the "primary reaction volume" (Chiu et al. 2005), as it dissolves the methoxide ion, which is the active species that promotes the transesterification reaction (Csernica and Hsu 2012a). At the beginning of the reaction, vigorous stirring leads to the rapid formation of an emulsion. At the same time, diglycerides begin to form and they function as phase transfer agents, helping the reaction to proceed. As it advances, a homogeneous phase forms, referred to by some authors as a single phase (Gunchavai et al. 2007). In this phase, the contact between methoxide and triglycerides is facilitated by providing a high conversion rate to the reaction. Somewhere in the reaction coordinate, a second pair of phases begin to form, FAME and unreacted oil as the upper phase and the glycerol/methanol mixture as the lower phase. This new two-phase formation does not mean the end of the reaction. The reaction continues to occur, but at a slower rate (Csernica and Hsu 2013b, Chiu 2005). This contributes to the separation between the reactive species—the remaining unreacted oil is less soluble in the new glycerol/methanol phase, where the concentration of the methoxide ion is higher leading to the decrease in speed reaction. Figure 2(a) shows the results of a set of batch biodiesel preparations (n = 13) after the

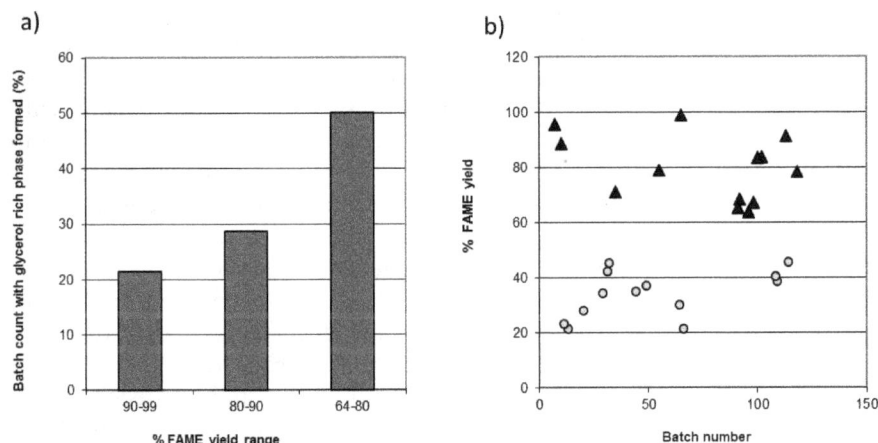

Figure 2. (a) FAME content (%) in a set of batches after 12 h reaction period, with the lower glycerol/methanol phase already formed. (b) FAME yield (%) after 12 h of reaction for a collection of 26 batches. When the FAME yield is greater than 65–67%, the reaction medium shows a two-phase system with glycerol/methanol as the lower phase, batches assigned with the symbol ▲. The batches marked with ○ were those in which the glycerol phase has not yet been formed.

normal reaction period (12 h), with the FAME content analyzed, in batches in which the lower phase of glycerol/methanol has already formed.

As can be seen, under the experimental conditions used, most of the lots have not yet reached the desired yield in FAME—greater than 96%, showing more modest conversion results in the 64–80% range. In general, actions are necessary to achieve the full reaction, or at least with yield values greater than 96%. Heating at 60°C for 3–4 h, under agitation, generally leads the reaction to the desired final yield. Theoretical studies on the reaction kinetics are published with the values of the formation and consumption rates of all species involved in the preparation of FAME. The use of these data is problematic under experimental conditions because there are several phase effects, which were not taken into account in these studies (Slinn and Kendall 2009).

The speed of conversion of triglycerides to FAME depends mainly on two factors:

 i) the degree of contact between methoxide and triglycerides and

 ii) the absence of water in the reaction medium.

The first is especially important when the reaction medium shows two phases with relatively different density values, that is, at the beginning and at the end of the reaction. To improve the phase transfer, actions such as manipulating the type and frequency of agitation, heating and adding phase transfer agents can be performed. Regarding the water content in the reaction medium, factors such as the quality of methanol, the drying of the oil that may contain emulsified water and the drying of the methanol recovered by distillation are factors that will have to be evaluated. Although the reaction rate depends on these aspects, the formation of the glycerol/methanol phase depends on the relative concentration of all the components of the reaction, namely the amount of alcohol added at the beginning of the reaction. To speed up the reaction, it is common to use an excess of alcohol, usually with a molar ratio of 1 to 6 or even higher, respectively for triglycerides (T) and methanol (M). From the relatively high number of batches to which this work refers, carried out under experimental conditions that have persisted for many years, detailed composition data of the reaction medium are known in the vicinity of the point at which the densest glycerol/methanol phase begins to form. These are shown in Figure 2(b) (n = 26), chosen from a set of FAME preparation batches. Most batches in which the glycerol/methanol phase has already occurred show FAME yields reached in the range of 65–67%. The lowest conversion yield recorded in a batch that already shows the glycerol/methanol phase shows a value of 63.8% in the FAME yield. Attempts to compare it with data from the literature were unsuccessful due to its non-existence. Even so, a brief reference was found for the occurrence of the formation of a lower phase of glycerol/methanol close to 70% yield in FAME, under the same stoichiometric conditions used in this work (Zhou et al. 2006). The knowledge of the exact point of formation of the glycerol/methanol phase and the complete characterization of the system components in it, such as the degree of conversion to FAME, would be very useful for the following reasons: when the phase of glycerol forms, the reaction speed slows down due to its characteristic of hindering phase transfer. This leads to the need of other steps for finishing the reaction, such as stirring operations, adding alcohol, and/or heating. Delaying the formation of the glycerol/ methanol phase would make it possible to maintain the reaction speed at a high value and achieve a high yield when converting to FAME. In addition, the formation of the glycerol/methanol phase can constitute a very simple way to monitor the degree of completeness of the reaction. In order to better understand the behavior of the phases, in the reaction of converting triglycerides to FAME, a mathematical model was constructed that was eventually capable of explaining the results observed experimentally and potentially inferring new information from it.

4. Construction of a mathematical model to determine the point of formation of the glycerol/methanol phase in the FAME preparation reaction coordinate

The miscibility of the different species involved in the FAME preparation reaction medium, from triglycerides and alcohols, determines the existence of immiscible phases. The exact point at which the lower glycerol/methanol phase forms depends on the amount of methanol used at the start of the reaction, usually an excess, in addition to the relative amounts of all other species present in the reaction medium. To fully understand the composition of each phase, it is necessary to know the inter-miscibility of all components of the reaction. Oil (triacylglycerols) and FAME are mutually miscible in all proportions, as are methanol and glycerol. Searching the literature for data on the solubility of methanol in FAME/oil mixtures did not result in any information. However, it is expected that some methanol will be solubilized in FAME/oil mixtures, and this solubility will increase with the FAME concentration. Glycerol is very insoluble in triacylglycerols and also in FAME (Csernica and Hsu 2012a). At the beginning of the reaction, the system is well defined with oil as the densest phase and methanol/catalyst as the upper phase. There follows the formation of a homogeneous intermediate phase, (Gunchavai et al. 2007) and in more advanced stages of the reaction, a lower phase of glycerol/methanol begins to form. An exact model must take into account the presence of mono- (MG) and diglycerides (DG) and must account for the excess molar volumes of the mixture, as these alter the density values. For MG and DG, a literature search on the partition coefficients in methanol/ oil mixtures revealed a lack of data. As its presence is transitory and more relevant at the beginning of the reaction, its contribution was not considered. Regarding the data of mixture excess molar volumes, they are described in the literature and have been evaluated. For binary mixtures of alkyl esters/ triacylglycerols, only slight deviations from ideality were found and the excess volume decreases with the increasing number of carbon atoms in the ester (Sazonov et al. 2002). Densities and excess molar volumes for the mixture of alcohols and glycerol also show slight differences from ideality (Li et al. 2007, Alkindi et al. 2008). The volume variation values described are very low, in the range of 0.3–1.0%, in absolute value of the molar volume. These data meant that in this approach the excess molar volume was not accounted for in the process of dissolving binary mixtures, considering them as ideal. With these restrictions, a simple model was built with two binary phases: the reactants, triacylglycerides and methanol (T + M) and the reaction products, biodiesel and glycerol (FAME + G). The relevant data for the characterization of the components are the molar mass and the density, which are summarized in Table 1. The values for these last variables are varied, based on the type of oil and respective fatty acid compositions. In this work, residual edible oil was known to be a mixture in which soybean oil was the main component, for other cases corrections must be made for the density and molar mass value (Nouredini et al. 1992).

Following this procedure, a spreadsheet was built to simulate the progress of the reaction coordinate, by successive increments, expressed in molar fraction, of the reagents in the products, respecting the two immiscible phases mentioned above. The same is shown in Table 1. At the beginning of the reaction (t_0) for any amount of oil and methanol in Kg, the molar fractions xi of oil x_T and methanol x_M were calculated, with Σ xi = 1. The reagents are converted, by small increments in the products (FAME + G), also expressed in molar fraction xi values, respecting the stoichiometric law. For each increment of the molar fraction, two groups of components formed based on mutual miscibility giving rise to two binary

Table 1. Phase composition model for the FAME preparation reaction coordinate.

| | | Molar fraction, xi | | | | FAME mass | Density | | | FAME convers. |
| | | Reactants | | Products | | | Phase 1 | Phase 2 | | |
it	incr	T	M	FAME	G	(g)	T+FAME	M+G	Sign	(% mass)
t0	0.003	0.250	0.750	0.000	0.000	0.00	0.9193	0.7914	1	0.0
1		0.247	0.741	0.009	0.003	0.28	0.9190	0.7948	1	0.1
2		0.244	0.732	0.018	0.006	0.55	0.9187	0.7982	1	0.3
3		0.241	0.723	0.027	0.009	0.83	0.9183	0.8017	1	0.4
//										
n		0.040	0.120	0.630	0.210	19.32	0.8778	0.8997	−1	54.5

Notes. n-Number of the iteration, Incr-value of the increment in each iteration (0.003), T-triglycerides, M-methanol, FAME-methyl esters of fatty acids; G = glycerol. Oil (soybean oil), average molar mass = 872 g, density = 0.9193 (Weast 1989); FAME molar mass = 297 g, density = 0.8739 (Andreatta et al. 2008); molar mass of glycerol = 92 g, density = 1.2613 (Sazonov et al. 2002); molar mass of methanol = 32 g, density = 0.7914. Phase 1 is a mixture of oil and FAME; phase 2 is a mixture of methanol and glycerol. The signal refers to the glycerol/methanol phase as the least dense phase (1) and (–1) as the denser phase.

phases (T + FAME, phase 1) and (M + G, phase 2). Assuming the molar volume in excess of the mixture equal to zero, the density values of each phase were calculated according to González (equation 3) (González et al. 2000).

$$0 = (x1\ M1 + x2\ M2)/\rho ph - [(x1\ M1/\rho 1) + (x2\ M2/\rho 2)] \tag{3}$$

where ρph is the phase density of the mixture and ρi is the density of each pure component. Mi is the molar mass of component i. In the spreadsheet, the comparison of the density of the two phases is made by the value of the sign of the difference between the values of each phase. The FAME mass produced in each increment was included to allow the determination of the reaction yield (expressed as a percentage of mass, shown in the rightmost column in Table 1).

The simulation using the model with different proportions of the initial molar ratio between the reactants—methanol and triglycerides—allows the calculation of the composition of the two phases of the system in which the formation of the densest glycerol/methanol phase occurs, together with the yield value of FAME at that point. The results obtained by the simulations are shown in the form of a graph, in Figure 3 (marked with ×). The dots roughly define a straight line, meaning that the formation of the glycerol/methanol

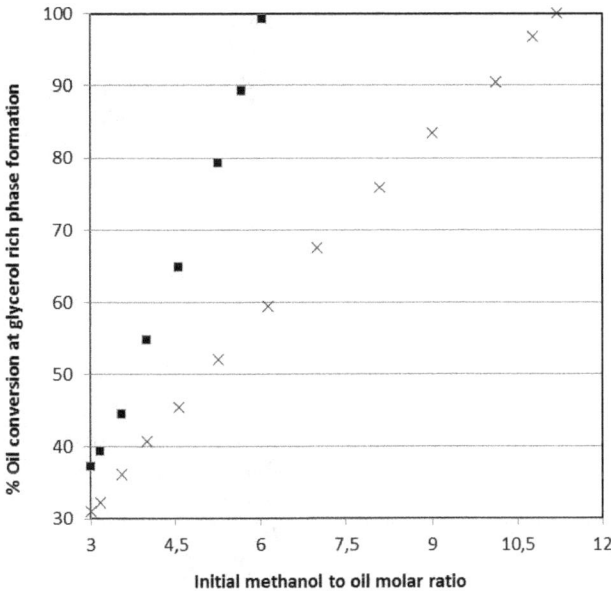

Figure 3. Graph of the calculated FAME content, at the point of the formation of the glycerol/methanol phase, in relation to the initial molar ratio of the reagents (marked with a ×), using the model considering two binary phases. Marked with ■, the same study is shown, but considering the phases as ternary, including the predictable methanol content in both phases. Experimentally, it is observed that under the conditions used, the formation of the glycerol/methanol phase (here referred as the glycerol rich phase) occurs when the conversion of the oil into FAME reaches a 65–67% yield.

phase depends linearly on the initial methanol/triglyceride molar ratio. The experimental conditions used the molar ratio of oil to methanol equal to 1:6 and for this value, the model indicates a FAME yield in the range of 55–57% for the formation of glycerol/methanol denser phase. The value is somewhat less than the experimentally observed value (65–67%).

Methanol has some solubility in both phases of the reaction medium and the first thought and attempt to improve the model was to take into account the concentration of methanol that is likely to be solubilized in each phase. Data for solubility are available in the literature, but only for methanol / FAME mixtures (Csernica and Hsu 2012, Alkindi et al. 2008, Andreatta et al. 2008). From these data, linear equations were calculated to determine the approximate methanol content in any of the phases, within the range of concentrations used. When taking into account the contribution of methanol, the results of the simulations of the formation of the glycerol/methanol phase are shown in the graph in Figure 3, marked with a ■. The results differ considerably from the observed experimental values. The effects of the presence of methanol in the two phases do not seem to explain the observed results. A detailed analysis of experimental data existing at various stages of the reaction and the behavior of the model, resulted in a potential identification of the source of the discrepancy. The glycerol/methanol phase in the model appears to assume higher density values in the initial stages of the reaction than those measured experimentally (data not shown). Attempts have been made to find the reason for this effect. In the model built, glycerol was increased as methanol is being consumed, but glycerol (G) is formed in a multi-step process from triglycerides, with diglycerides (DG) and monoglycerides (MG) as intermediates:

step	1	2	3
	T →	DG →	MG → G
Glycerol			
formation	no	no	yes

Steps 1 and 2 do not give rise to any glycerol, which is formed only in step 3. This means that, regardless of the reaction speed, which can assume different values depending on the condition of the reagents, in the initial steps the main products of the reaction are the diglycerides (DG), subsequently these are converted into monoglycerides (MG) and only when the concentration of the latter reaches high values, the formation of glycerol reaches significant values, reaching the total yield in the final stages of the reaction (step 3). The delay in the formation of highly dense glycerol, which induces the formation of the final lower phase, is caused by the mechanism in stages. A strategy was designed to derive an approximate mathematical expression for the formation of glycerol, in order to correct the model. The strategy was to roughly calculate the probability of one, two and three transesterifications of the acyl bonds by adding 3 methanol molecules to 9 possible ester bonds of three triglyceride molecules (expressed in molar fraction, with 1/3 mol of methanol added for the complete reaction). With this probability calculated, in a second step the values obtained allow the calculation of the molar fraction values for DG, MG and G, after adding 2/3 of the number of moles of methanol necessary to effect the total transesterification of the triglycerides. This procedure is presented in the form of a diagram in Figure 4(a); see the notes for the explanation of the notation used.

a) 3 M + 3 T (OOO OOO OOO), 9 acyl positions possible to transesterify (⊕); see notes.

$\binom{9}{3}$ = 9!/3!(9-3)!= 84 possible combinations

Brief analysis:

⊕⊕⊕ OOO OOO, ...   3 in 84, p= 3/84≈0.04

⊕OO ⊕⊕O OOO, ...   24 in 84; p= 24/84=0.29 ≈ 0.3

⊕OO ⊕OO ⊕OO, ...   57 in 84; p= 57/84=0.68 ≈ 0.7

b) (i) Beginning

OOO OOO OOO OOO OOO OOO OOO OOO OOO OOO (10T)

(ii) With 1/3 third mol methanol spent

⊕OO ⊕OO ⊕OO ⊕OO ⊕OO ⊕OO ⊕OO |⊕OO ⊕⊕O OOO   (8DG+1MG+1T)
        p≈0.7 | p≈0.3   most probable

(iii) with more 1/3 third mol methanol spent, totaling 2/3 moles.

⊕⊕O ⊕⊕O ⊕⊕O ⊕⊕O ⊕⊕O ⊕⊕O ⊕⊕O |⊕⊕⊕ ⊕⊕⊕ OOO   (7MG+2G+1T)
        p≈0.7 | p≈0.3   or
⊕⊕O ⊕⊕⊕ ⊕OO| ⊕⊕O ⊕⊕O ⊕⊕O ⊕⊕O ⊕⊕O ⊕⊕⊕ ⊕OO   (6MG+2DG+2G); ...
    p≈0.3 | p≈0.7

(iv) With full methanol spent

⊕⊕⊕ ⊕⊕⊕ ⊕⊕⊕ ⊕⊕⊕ ⊕⊕⊕ ⊕⊕⊕ ⊕⊕⊕ ⊕⊕⊕ ⊕⊕⊕ ⊕⊕⊕   (10G)

Notes. OOO-triglyceride (T), ⊕OO-diglyceride (DG), ⊕⊕O-monoglyceride (MG), ⊕⊕⊕-glycerol (G), ⊕acyl bond break; M=methanol,

Figure 4. (a) Probability of breaking the acyl bonds when adding 3 methanol molecules to three triglyceride molecules (9 acyl bonds). (b) Application of the break probability values obtained in (a) to 10 triglyceride molecules (T) after 0 (i), 1/3 (ii), 2/3 (iii) and full of the number of moles of methanol needed for the complete conversion of triglycerides to FAME.

Suppose 3 molecules of methanol (M) are reacted with 3 molecules of triglycerides (T). Assuming that any of the acyl positions of the triglycerides are equally likely to undergo transesterification, the distribution of the probability of formation of the break products of the acyl bond can be deduced from calculating the number of combinations of 3 objects in nine possible positions. There are 9!/3! (9-3)! = 84 different ways to break the 9 acyl bonds present in 3 triglyceride molecules, with 3 methanol molecules (in the form of methoxide). Careful analysis shows that of these 84 possibilities, only three lead to the breakdown of the acyl bond in the same triglyceride (probability, p = 3/84 = 0.04; see Figure 4); 24 generate a DG, an MG and an unreacted triglyceride molecule (p = 24/84 = 0.28) and the remaining 57 generate 3 DG (p = 57/84 = 0.67). To determine the triglyceride break probability values in terms of MG, DG and G formation, a simulation with 10 triglyceride molecules is shown in Figure 4(b). For simplicity, the formation of FAME and glycerol was considered in four stages, at the beginning (t_0) (i), with 1/3 (ii), 2/3 (iii) and the totality (iv) of the stoichiometric molar quantity of methanol added to allow the completion of the reaction. These intermediate probability values will give, after a curve fit, a mathematical equation for the formation of glycerol in the course of the reaction coordinate. In Figure 4(b) from (i) to (iv): In the beginning, there were only 10 triglyceride molecules (T); the most likely reaction result of the first molar third of methanol added to the triglyceride is identified in Figure 4(b) (ii) and is 8DG + 1MG + 1T (an unreacted triglyceride). With this distribution, the formation of glycerol will be very close to zero and will not exceed a value of p = 3/84 ≈ 0.04, as calculated in Figure 4(a). Applying the probability of bond breakage to (ii), with an additional 1/3 mole of methanol added to the medium, the result of the likely reaction is the formation of 7MG + 2G + T (unreacted triglyceride) as shown in Figure 4(b) (iii). It is easily seen that several other combinations are possible, such as 6MG + 2DG + 2G, also shown in Figure 4(b) (iii). However, in terms of glycerol formation, the highest probability will be, in any case, 2G. Thus, the greatest probability of producing glycerol after 2/3 moles of methanol added in transesterification will be 2/10 = 0.20. With the remaining methanol added, a glycerol formation will be complete (iv).

By placing the values of probability of glycerol formation versus the number of moles of methanol added in the complete transesterification reaction of the triglycerides, the graph shown in Figure 5(a) is obtained. As noted, the formation of glycerol is very low in the first 2/3 moles of methanol added, with its maximum formation at the end 1/3 moles of methanol added. This behavior generates an exponential type curve and is originated by the multi-step glycerol formation mechanism shown in Figure 4. Adjusting the points with exponential and polynomial curves, the best fit to the points was obtained with a 3rd degree polynomial. From this curve adjustment, the mathematical equation is obtained for the probability of formation of glycerol in the reaction coordinate, here called pG, shown in the graph of Figure 5(a).

a)

b)

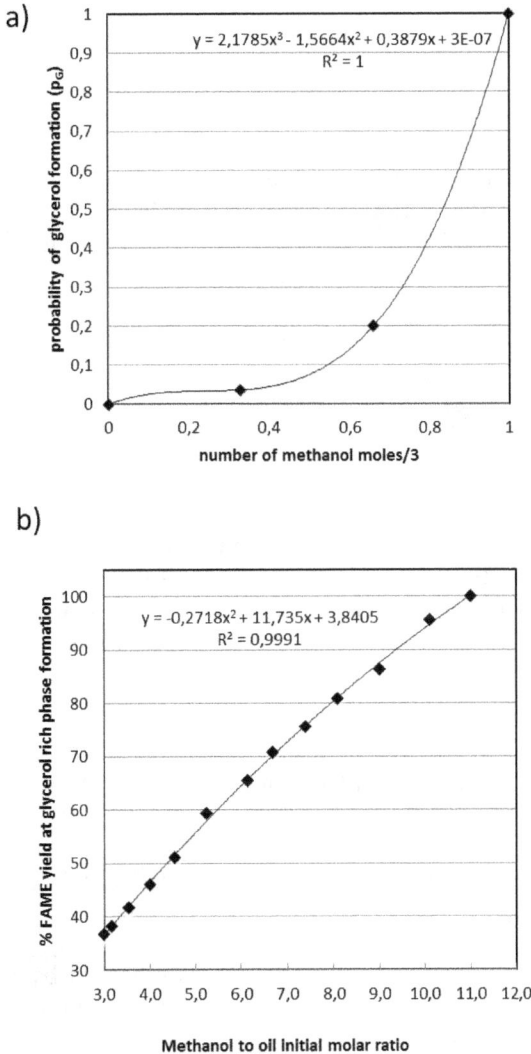

Figure 5. (a) Probability of glycerol formation in the conversion of triglycerides to FAME, after 0, 1/3, 2/3 of the required molar methanol added. A 3rd degree polynomial curve was fitted to the points. (b) Graph of the updated model of the glycerol/methanol phase formation point versus the initial methanol molar fraction used in the preparation of FAME (fitted with a degree polynomial curve), which includes the correction for the delayed glycerol formation shown in (a). The results observed between the calculated and experimental values show to be in agreement for the formation of the glycerol/methanol phase.

The correction of the spreadsheet of the forecast model was made by including columns for the content and DG and MG and correction of the molar fraction of glycerol (xGC). Thus, the molar fraction of glycerol corresponds to the product of the probability of its formation (pG) multiplied by the molar fraction gradually increased in the spreadsheet (xGC= pG. xG). The introduction of this parameter also causes the reaction coordinate to be normalized, in relation to its completeness.

The results of the corrected model for the glycerol/methanol phase formation are shown in the graph in Figure 5(b). As can be seen in the Figure, an initial ratio of reagents (T and M) of 1: 6 results in the prediction of occurrence for the formation of the glycerol/methanol phase at values near to 64–65% in FAME yield, values very close to that found in the experimental results (65–67%). This was the desired result, indicating a satisfactory model response in the correspondence between the calculated and the observed value.

Relevant information can be inferred by analyzing the model results for the phase behavior in different situations, including:

(i) with the use of reagents (T/M) at the beginning of the reaction with a molar ratio of 1: 3, the formation of the glycerol/ methanol phase occurs early in the reaction coordinate, in the FAME yield range of about 35–37%;

(ii) with the use of an initial molar ratio of the reagents greater than about 1:12 for oil/methanol (T/M), the formation of a glycerol/methanol phase that is denser than the FAME/oil phase should not occur, remaining throughout the reaction coordinate methanol and glycerol in the upper phase.

These results can be used to slow down the deceleration of the FAME production rate that occurs after the formation of the glycerol/methanol phase, however the use of a greater amount of methanol increases the cost for its recovery and may hinder the biodiesel recovery process.

5. Experimental verification of the results of the glycerol/ methanol phase formation identification model

In order to experimentally verify the results obtained by the model built above under experimentally unverified conditions, a new set of laboratory tests for the preparation of FAME from soybean oil were carried out, using the reagents at the beginning of the reaction in different molar proportions. The experimental conditions were kept close to those of the pilot unit from which the analytical data were collected, the main differences being the use of new soybean oil, the reaction at a temperature of 60°C and the constant magnetic stirring of the reaction mixture, with brief time intervals for separation and identification of the ordering of the phases. These conditions allow the preparation of FAME within shorter periods, which was fully verified by the parallel HPLC analysis of the reaction product. The results of this study are shown in Figure 6.

The results show the formation of a lower glycerol/methanol phase, as expected, using molar ratios of oil/methanol equal to 1:3 to 1:12. However, using a 1:15 ratio, the glycerol/methanol phase occupies the upper position, showing its characteristic brown color. The result observed in tube 5

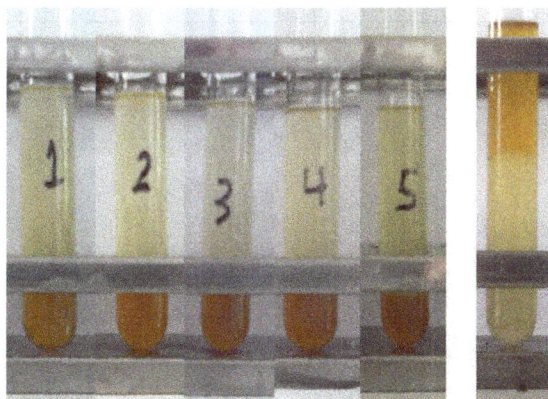

Figure 6. Monitoring the behavior of the phases in complete FAME preparation tests using soybean oil/methanol as reagents with 0.5% (w/v) KOH catalyst with initial molar ratios of oil/methanol equal to 1:3 (1); 1:5 (2); 1:8 (3); 1:10 (4); 1:12 (5) and 1:15 (the rightmost tube). The latter, shows the glycerol/methanol phase as being the less dense upper phase (the dimensions of the tubes are different).

(1:12 molar oil/methanol proportions) is very close to the inversion point indicated by the model, but this has not yet shown the phase inversion. Slight differences in oil density may be responsible for the observed difference.

The detailed analysis of the model results adds an understanding of the density variation of the two phases involved in the reaction coordinate of biodiesel preparation (FAME) from edible oils and methanol. The homogeneous transitional phase, sometimes referred to as a single phase, does not apply to this discussion, as it, if any, is transient and does not interfere with the initial or final conditions in which the formation of two phases always occurs. For this reason, the model described here uses different initial molar ratios for the oil and methanol reagents, considering the presence of a biphasic binary system during the entire course of the reaction. Figure 7 shows the density values, calculated from the model, of the glycerol/methanol phase in the reaction coordinate with different initial oil/methanol proportions.

The results indicate that the density of the oil/FAME phase varies only slightly from the beginning to the end of the reaction, decreasing from values from 0.898 to 0.874, and these are almost independent of the initial molar proportions for triglycerides (T) and FAME. This is due to the similarity in the density of the reagent triglycerides or oil (T) and FAME reaction products, mutually miscible. However, the glycerol/methanol phase undergoes a much more pronounced density variation from the beginning to the end of the reaction. Using a 1:11 oil/methanol molar ratio, the glycerol/methanol phase density increases from 0.791 to about 0.860 at the end of the reaction. With an initial molar ratio of oil/methanol equal to 1:6, the final density value increases to about 0.998 and with the initial molar ratio equal to 1:3, the phase density increases even more, to values closer to the density glycerol, such as

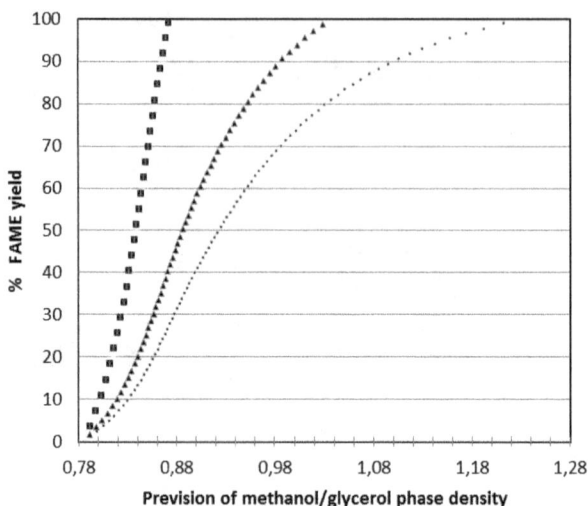

Figure 7. Prediction of the density of the glycerol/methanol phase in the coordinate of the reaction of preparation of FAME from soybean oil and methanol, for initial molar ratio for the oil/methanol pair equal to 1:3 (•), 1:6 (▲) and 1:11 (×).

1,200 (see Figure 7). This is due to the relatively high value of the glycerol density that forms mainly in the final stages of the reaction. The lower the methanol used at the beginning of the reaction, the greater the variation in density of the glycerol/methanol phase will be observed.

The results described above point out that the measurement of the density value of the glycerol/methanol phase may constitute a sensitive and simple method in determining the degree of conversion of oils into FAME, namely in the situation of the final phase already being formed. This data can be used to allow conversion monitoring in small biodiesel production units, usually without access to more elaborate analytical techniques such as GC and HPLC, with an instrumental technique accessible to anyone— the measurement of density. Table 2 shows the density values obtained by the model described here, corresponding to the most probable FAME yields from soybean oil, obtained by measuring the density of the glycerol/methanol phase, using different initial proportion of reagents.

When comparing a set of ten batches of FAME preparation, under the standard conditions used in the present work, the yield obtained by the method of measuring the density of the glycerol/methanol phase presents a difference of 5.65 ± 2.3 when compared with the evaluation of the FAME yield of the same samples by the HPLC method mentioned above. This result demonstrates that this simple analytical method can be used for those who prepare biodiesel and do not have access to more accurate analytical techniques, such as GC or HPLC, to determine the degree of conversion of oils or fats into FAME. Table 2 shows the expected density values of the glycerol/methanol phase, using different initial proportions of the reagents, in relation to the conversion yield into FAME.

Table 2. Approximate yield of FAME in the conversion of soybean oil, obtained from the calculated value of the density of the glycerol/methanol phase (lower phase), for reactions with different initial proportions of oil/methanol.

Density of Glycerol rich phase	Initial molar proportion of triglycerides/methanol			Density of Glycerol rich phase	Initial molar proportion of triglycerides/methanol		
	1:03	1:06	1:10		1:03	1:06	1:10
	App. FAME % yield				App. FAME % yield		
0.780	1.2	3.1	5.7	1.020	78.2	98.3	
0.785	1.0	2.8	4.5	1.025	79.1	99.5	
0.790	1.1	2.9	4.6	1.035	80.9	100.0	
0.795	1.5	3.5	5.9	1.040	81.7		
0.800	2.2	4.4	8.3	1.045	82.5		
0.805	3.0	5.7	11.7	1.050	83.3		
0.810	4.1	7.3	16.0	1.055	84.0		
0.815	5.3	9.2	21.0	1.060	84.7		
0.820	6.8	11.4	26.8	1.065	85.4		
0.825	8.3	13.8	33.1	1.070	86.1		
0.830	10.0	16.3	39.8	1.075	86.7		
0.835	11.8	19.1	47.0	1.080	87.3		
0.840	13.7	21.9	54.3	1.085	88.0		
0.845	15.7	24.9	61.8	1.090	88.6		
0.850	17.8	28.0	69.4	1.095	89.2		
0.855	19.9	31.1	76.8	1.100	89.8		
0.860	22.1	34.2	84.1	1.105	90.4		
0.865	24.3	37.4	91.1	1.110	90.9		
0.870	26.5	40.6	97.7	1.115	91.5		
0.875	28.8	43.8	100.0	1.120	92.1		
0.880	31.0	46.9		1.125	92.7		
0.885	33.3	50.0		1.130	93.3		
0.890	35.5	53.0		1.135	93.8		
0.895	37.7	56.0		1.140	94.4		
0.900	39.9	58.8		1.145	94.9		
0.905	42.1	61.6		1.150	95.5		
0.910	44.3	64.3		1.155	96.0		
0.915	46.4	66.9		1.160	96.6		
0.920	48.4	69.3		1.165	97.1		
0.925	50.5	71.6		1.170	97.6		
0.930	52.4	73.9		1.175	98.1		

Table 2 Contd. ...

...Table 2 Contd.

Density of Glycerol rich phase	Initial molar proportion of triglycerides/methanol			Density of Glycerol rich phase	Initial molar proportion of triglycerides/methanol		
	1:03	1:06	1:10		1:03	1:06	1:10
	App. FAME % yield				App. FAME % yield		
0.935	54.3	76.0		1.180	98.6		
0.940	56.2	77.9		1.185	99.0		
0.945	58.0	79.8		1.190	99.4		
0.950	59.7	81.5		1.195	99.8		
0.955	61.4	83.1		1.200	100.0		
0.960	63.1	84.6					
0.965	64.6	86.1					
0.970	66.1	87.4					
0.975	67.6	88.6					
0.980	69.0	89.8					
0.985	70.3	90.9					
0.990	71.6	92.0					
0.995	72.8	93.0					
1.000	74.0	94.0					
1.005	75.1	95.0					
1.010	76.2	96.1					
1.015	77.2	97.1					

Acknowledgment

The author would like to thank all teachers, non-teachers, technicians, and students of ESAC/IPC who contributed so that its pilot biodiesel preparation unit could be assembled, operated, and maintained in good conditions over the years.

6. References

Alkindi, A. S., Al-Wahaibi, Y. M. and A. H. Muggeridge. 2008. Physical properties (Density, excess molar volume, viscosity, surface tension, and refractive index) of ethanol + glycerol. J. Chem. Eng. Data. 53: 2793–2796.

Andreatta, A. E., Casás, L. M., Hegel, P., Bottini, S. B. and E. A. Brignole. 2008. Phase equilibria in ternary mixtures of methyl oleate, glycerol, and methanol. Ind. Eng. Chem. Res. 47: 5157–5164.

Assuncao, M. F. G., Varejao, J. M. T. B. and L. M. A. Santos. 2017. Nutritional characterization of the microalga Ruttnera lamellosa compared to Porphyridium purpureum. Algal Research-Biomass Biofuels and Bioproducts 26: 8–14.

Canakci, M., Erdil, A. and E. Arcaklioglu. 2006. Perfomance and exaust emissions of a biodiesel engine. Apll. Energ. 82: 594–605.

Chiu, C. W., Goff, M. J. and G. J. Suppes. 2005. Distribution of methanol and catalysts between biodiesel and glycerin phases. AIChE Journal 51: 1274–1278.

Csernica, S. N. and J. T. Hsu. 2012. The phase behavior effect on the kinetics of transesterification reactions for biodiesel production. Ind. Eng. Chem. Res. 51: 6340–6349.

Csernica, S. N. and J.T. Hsu. 2013. Inhibitory effect of the byproduct glycerol resulting from the phase behavior on the transesterification reaction kinetics. Energy Fuels 27: 2167–2172.

Daroch, M., Geng, S. and G. Wang. 2013. Recent advances in liquid biofuel production from algae feedstocks. Apll. Energ. 102: 1371–1381.

EU European Committee for Standardization. 1991. Olein content determination in olive oil. European Community Official Journal, NL 248/29.

González, C., Resa, J. M. and J. Lanz. 2000. Excess volumes of binary mixtures that contain olive oil with alkyl and vinyl acetates. J. Oil & Fat Ind. 77: 985–990.

Gunchavai, K., Hassan, M. G., Shama, G. and K. Hellgardt. 2007. A new solubility model to describe biodiesel formation kinetics. Trans. IChemE 85: 383–389.

Hill, J., Nelson, E., Tilman, D., Polasky, S. and D. Tiffany. 2006. Enviromental, economic, and energetic costs and benefits of biodiesel and ethanol biofuels. PNAS 103: 11206–11210.

Leung, D. Y. C., Wu, X. and M. K. H. Leung. 2010. A review on biodiesel production using catalysed transesterification. Apll. Energ. 87: 1083–1095.

Li, Q. S., Su, M. G. and S. Wang. 2007. Densities and excess molar volumes for binary glycerol +1-Propanol, +2-Propanol, +1,2-Propanediol, and +1,3-Propanediol mixtures at different temperature. J. Chem. Eng. Data 52: 1141–1145.

Martins, C. B., Varejao, J. M. T. B. and L. M. A. Santos. 2016. Biotechnological potential of Haematococcus pluvialis Flotow ACOI 3380. Nova Hedwigia 103: 547–559.

Meher, L. C., Vidya Sagar, D. and S. N. Naik. 2006. Tecnhical aspects of biodiesel production by transesterification—a review. Renew. Sustain Energy Rev. 10: 248–268.

Miller, R. G. and S. R. Sorrell. 2014. The future of oil supply. Phil. Trans. R. Soc. A 372: 20130179.

Nouredini, H., Teoh, B. C. and L. D. Clements. 1992. Densities of vegetable oils and fatty acids. J. Am. Off. Chem. Soc. 69: 1184–1188.

Sazonov, V. P., Sazonov, N. V. and N. I. Lisov. 2002. Quaternary system nitromethane +1-Hexanol +Octanoic acid +1,2,3-Propanetriol with three liquid phases. J. Chem. Eng. Data. 47: 1462–46.

Slinn, M. and K. Kendall. 2009. Developing the reaction kinetics for a biodiesel reactor. Biores. Technol. 100: 2324–2327.

Talebian-Kiakalaieh, A., Amin, N. A. S. and H. Mazaheri. 2013. A review on novel processes of biodiesel production from waste cooking oil. Apll. Energ. 104: 683–710.

Varejão, J. M. T. B. and M. C. Cruz Costa. 2015. New on-column GC sample injection techniques: Making holes in separation capillaries. Scientia Chromatographica 6: 277–282.

Varejão, J. M. T. B. and R. Nazaré. 2017. Ethanol production from macroalgae biomass. Chapter 6. *In*: Leonel Pereira (ed.). Algal Biofuels CRC Press, Boca Raton. USA.

Vicente, G., Martínez, M. and J. Aracil. 2004. Integrated biodiesel production: A comparison of diferent homogeneous catalysts systems. Biores. Technol. 92: 297–305.

Weast, R. C. 1989. Handbook of Chemistry and Physics, 69th Edition. Boca Raton. USA.

Zhou, H., Lu, H. and B. Liang. 2006. Solubility of multicomponent systems in the biodiesel production by transesterification of Jatropha curcas L. oil with methanol. J. Chem. Eng. 51: 1130–35.

Chapter **9**

Preparation and use of Biodiesel in a Continuous Process using Alcohol/Water Mixtures

1. Introduction

The residues of edible oils are a by-product of different types of residual animal and vegetable fat that tends to grow in terms of volume in the food chain. They are, materials from biomass that are generated in parallel with the growth of the world population and are associated with a highest standard of well-being. Frying food is a food preparation technique that is appreciated all over the world due to its characteristics of giving food flavor, texture and color, while guaranteeing food safety in relation to the presence of pathogenic microorganisms (Multari et al. 2019). The production of biomass with a high lipid content, such as oilseeds, microalgae and animals has steadily increased in recent years due to market demand, which in turn will generate a large volume of biomass residues rich in lipid (Simsek et al. 2020).

The recovery of these lipids and their conversion into the corresponding esters by means of the base-catalyzed transesterification reaction with alcohols produces alkyl esters of fatty acids such as FAME, which can be incorporated into diesel fuel with advantages for the operation of the engine and reduced emissions of coal particles (Mujtaba et al. 2020), is one of the few possibilities for valuing these materials. The nature of the processes that use oils and fats gives their residues the presence of several types of substances, the most common being water that forms emulsions with them, and suspended solids derived from the specific processing of fats, and free fatty acids (FFA).

In samples of lipid such as olive oil of inferior quality, and lipid obtained by extraction of some plant species, the content of free fatty acids (FFA) can reach values as high as 17%. In this case, its conversion by the classic transesterification reaction catalyzed by strong base (e.g., NaOH) becomes unfeasible, as the base converts the FFA in soap which emulsify the fats giving rise to solid materials. The transesterification reaction to work in good conditions requires less than 1% lipid FFA content. For this type of lipid, its conversion to FAME requires a preliminary esterification reaction, made with alcohols such as methanol in the presence of a small concentration of sulfuric acid (1–4% v/v) at a temperature of 60°C for 6 to 8 h. These conditions efficiently convert the FFA fraction into the corresponding ester. After this step, the subsequent transesterification of the triglycerides, which do not react under esterification conditions, is carried out with catalysis by strong base, in a slight excess to neutralize the acid and by adding the required amount of methanol, for example a triglyceride/methanol equal to 75/25 (v/v). Different conditions and catalysts may operate with many bibliographic reviews on the subject in the literature (Wong et al. 2019).

The separation of suspended solids in waste oils can be achieved by means of gravitational sedimentation for periods of time between 1–3 weeks, until a yellow or brown transparent oil is obtained. In the most difficult cases, the addition of 10–25% FAME may assist sedimentation. The removal of emulsified water in residual lipid must be carried out at a concentration that does not interfere with the transesterification reaction, usually in a content equal to or less than 100–150 ppm. The addition of CaO followed by filtration or pumping through a filter medium may be an appropriate treatment. The water content can originate from methanol, as there is a failure in the preparation of FAME when changing methanol suppliers, since not every methanol producer complies with the water content equal to that stated on the label; or by adding methanol recovered by distillate from the residual fraction of glycerol, the main by-product of the biodiesel preparation and which is usually obtained with a high water content—a by-product of the transesterification reaction. In other cases, the water is emulsified in the residual lipid in different forms—a common example in suspended and transparent hydrated substances. Other possible sources of contamination are leaks in washing water lines, contact with rainwater, air humidity, etc. In some of these cases, water may appear in the reaction medium due to the lack of tightness of the reactors.

One of the purposes of the preparation and use of biodiesel is to reduce the emissions of carbon particles originating in the diesel engine with the concomitant reuse of a residual material that does not have a visible end use. Diesel-powered cars will end in the near future, however diesel will continue to be used for many years in heavy transport, whether by land or sea, and eventually on airplanes.

Biodiesel is not part of the fossil hydrocarbon group, therefore, its burning does not contribute to carbon dioxide emissions, which are responsible for the ongoing effects of climate change. Methanol used in the preparation of biodiesel, on the other hand, is bought at the lowest price from major international producers. The main source used in its preparation involves the conversion of fossil coal into syngas, a mixture of hydrogen and carbon monoxide. This mixture, after the appropriate stoichiometric adjustment of the ratio of carbon monoxide (CO) to hydrogen (H_2), is compressed at low pressures (< 5 bar) in the presence of a microparticle copper catalyst (such as CuO_2) to be converted into methanol. The price of methanol depends heavily on its purity, water is the most common main contaminator, usually present in more than 1000 ppm. To fulfill the environmental premises of the use of biodiesel, it is desirable that the fossil methanol used in the transesterification reaction is avoided and replaced by a more ecological substitute, the obvious choice is ethanol. Fatty acid ethanol esters (FAEE) have properties similar to those prepared with methanol, and the change does not cause problems regarding the intended end use, which is burning in diesel engines. Ethanol is prepared worldwide from biomass, mainly in Brazil and the USA, from the fermentation of sugarcane molasses and the fermentation of low quality cereal starch, respectively. In Europe, cheap ethanol is a common by-product of the extensive wine-making process, from which residual pomace is subjected to a distillation process to obtain ethanol/water mixtures. The concentration of ethanol is very variable and is usually found in the range of 40–80% (v/v).

Thus, it would be desirable in the preparation of biodiesel to change the use of methanol for ethanol, the main problem being to solve the existence of high water content. Its removal, although possible, will add to the cost, making the final price of biodiesel uncompetitive. Simple distillation is an ineffective process for the purification of ethanol, since it forms an azeotropic mixture with water with a composition of approximately 96/4% (v/v) in ethanol/water. Physical or chemical methods allow to reduce or even eliminate the water content, such as the use of molecular sieves or the addition of dehydrating substances, such as calcium oxide (CaO). With these treatments the water concentration can be reduced to values below 1%, but with the problems mentioned above. In many cases, even this water concentration is unacceptable for the transesterification reaction, making additional ethanol drying techniques necessary. This would lead to the addition of other steps with energy costs and expenses, to a product that already has a small profit margin. This means that switching from methanol to ethanol is far from an easy process, at least with classic transesterification methods catalyzed by a strong base. Alternative processes compatible with the presence of water are necessary to carry out the esterification and transesterification reactions.

2. Effects of the presence of water on the triglyceride transesterification reaction

The preparation of FAME from triglycerides can be done effectively in a single step, using as a catalyst the strong base sodium methoxide ($CH_3O–Na^+$), in the range of concentrations equal to 1–3 M in methanol, in reaction times for total conversion as low as 10 min, if under strong agitation. These conditions are not suitable for processing large volumes of triglycerides due to the cost of sodium methoxide, usually prepared from the reaction of methanol with metallic sodium.

Sodium methoxide ($CH_3O–Na^+$) can also be prepared from methanol solutions, in low concentration, with the simple addition of a strong base, such as NaOH, KOH and Ca(OH)2. The balance that leads to the formation of sodium methoxide is shown in Equation (1) below.

$$CH3OH + KOH \quad <-> CH3O–K+ \; + \; H2O \tag{1}$$

$$CH3O–K+ \; + \; RCO2R' \quad \rightarrow \quad RCO2CH3 \;\; + R'O–K+ \tag{2}$$

$$R'O–K+ \; + \; H2O \;\; -> \;\; R'OH \; + \; KOH \tag{3}$$

Equation (1) is a balance with the by-product being water. If by any means sodium methoxide can be removed from the medium, the balance moves towards the products (Le Chatelier's law) and more sodium methoxide can be prepared. In the presence of triglycerides, the methoxide is quickly consumed when carrying out the transesterification reaction, producing FAME and the alkoxide (Equation 2), and thus continuing the equilibrium (1). Alkoxide in the presence of water produces alcohol and regenerates the base (Equation 3). Thus, Equations 1–3 exemplify the transesterification of a simple ester (RCO2R') with methanol, the same process occurs using triglycerides.

The presence of water in the reaction medium results in a drastic decrease in the reaction speed (2), that is, it shows a strong inhibitory effect. Since Equation (1) is an equilibrium; the presence of water shifts the balance to the left, decreasing the concentration of potassium methoxide (CH3O–K+), which is the species active in the conversion of ester (or triglyceride) in transesterification (Equation 2) and, therefore, in production rate of FAME. Hence the harmful consequences of the presence of water in the reaction medium for preparing biodiesel. To speed up the reaction, it is common to use an excess of alcohol, usually in the molar ratio of 1 to 6 or even higher, respectively for triglycerides and methanol. In the practice of producing biodiesel in a pilot plant, in batch mode, for years, it is known that to avoid problems of not being able to convert oils into FAME, the water level must remain below 100–150 ppm. With a high water content in reaction solution do not yield any product or do so at a very slow rate reaching visible concentration only for prolonged reaction times (example of conversion of 14% after one week of reaction), even under heating (60°C), and in the presence of strong agitation.

3. Preparation of FAME using enzymes

A possible alternative to promote the transesterification of triglycerides with alcohol/water mixtures is the use of enzymes such as lipases (Imanparast et al. 2019, Muanruksa et al. 2020), they are very efficient in aqueous media and do not require an increase in temperature. The general limitation of the use of these is the probable denaturation by high concentrations of alcohols, eventually making the process ineffective. Most of the known lipases undergo denaturation in aqueous solution with an alcohol content greater than 5–6% (Gihaz et al. 2018). Ways to work around this problem is through the use of mixed solvents or the immobilization of enzymes in polymeric meshes, both are referred to in the literature and will be able to successfully implement transesterification with alcohols in aqueous media (Sharma et al. 2019). An example is the use of t-butanol as a co-solvent, and the lipase catalyst naturally immobilized in biomass, as in the rice plant contaminated with the fungus *Aspergillus oryzae*, by simple grinding and drying (Wijaya et al. 2020). These catalysts are especially useful in systems designed to use alcohols in low concentrations. The problem associated with these systems is that the catalysts are expensive, the co-solvents require additional separation operations, adding complexity and cost, when compared to the simpler process of catalysis by strong base.

4. Preparation of FAME by continuous process with strong base catalysis and microwave irradiation

The use of heating by microwave radiation in promoting chemical reactions has been the subject of study and better knowledge in recent years, and its use is recognized for its catalytic effects in different types of reactions, which otherwise result in low performance (Nayak et al. 2019). The main reason is that microwave heating, unlike thermal heating, has the ability to activate excited rotational states (Bren et al. 2010), which transfer energy to vibrational modes and in subsequent processes for translational energy, with the final heating effect of the substance. The groups that absorb radiation and microwaves are those that have a dipole moment, that is, groups with some degree of polarity. The reactivity of some specific functional groups in excited vibrational states, such as the carbonyl groups involved in the formation of carboxylic acids and esters, show increased reactivity, resulting in activation. Reactions such as esterification and transesterification involve additions to the carbonyl group, a group with a dipole moment that causes it to absorb microwave radiation. In fact, esterification and transesterification reactions in the presence of a base and alcohol are more efficient and faster when done under microwave irradiation (Binnal et al. 2020). They are still tolerant to the presence of water. Phosphonic acid esterification reactions can be done easily with the use of microwave heating (Keglevich et al. 2012). The literature is full of examples of promoting esterification reactions (Davies et al. 2020), transesterification (Binal et al. 2020), with homogeneous or heterogeneous catalyst (Abbas et al. 2020, Barbosa et al. 2019, de Oliveira et al. 2020) and even without any catalyst (Nguyen et al. 2020), provided they are made with microwave irradiation.

In the production of biodiesel (Khedri et al. 2019), it would be convenient to change the use of methanol to ethanol, in the form of 96% azeotrope, a low-cost version and available globally as a by-product of fermented beverage production. The presence of 4% (v/v) of water completely inhibits the transesterification reaction with catalysis with strong bases, such as KOH or NaOH in concentration in the range of 1–4% (w/v).

In an attempt to use ethanol in the production of biodiesel, tests in a continuous flow process assisted by microwaves were attempted with different catalysts and experimental conditions (Ali et al. 2020, Mazo and

Rios 2010), but in many cases the yields are low and the reaction times exceed reasonable periods of time (Lee et al. 2018). Surprisingly, few studies have been done to assess the effects of water on the efficiency of the process. The possibility of preparing biodiesel in the presence of water would make the direct conversion of biological tissue and residues from triglycerides into FAME viable, without the need to dry the biomass. The fatty acid esters that are beginning to be produced are good extraction solvents, catalyzing the ongoing process of mobilizing the most inaccessible triacylglycerols. This process can make the production of FAME in a single stage viable from substrates such as microalgae, different types of plant and animal biomass residues, such as those accumulated in olive oil production, etc. The process can be assisted with ultrasound irradiation to facilitate cell disruption (Patel et al. 2019). Microalgae biomass is particularly suitable for this process, since they can be obtained in liquid suspension, and are capable of being pumped in normal conditions of a continuous reactor.

For these reasons it is highly recommended to develop non-enzymatic methods capable of promoting both the esterification reactions of free fatty acids and the transesterification of triglycerides with alcohol/water mixtures, as a way to optimize and in many cases make it possible to obtain biodiesel from the growing volume of lipid-rich waste materials.

The best knowledge of the behavior of the phases in the reaction coordinate in the biodiesel preparation (Chapter 8) can be used with advantage in the design of continuous systems, more efficient in the production of biodiesel.

Biodiesel production tests using mixtures of water/ethanol and water/methanol have been successfully implemented in continuous processes using direct microwave irradiation. An example is described, with the process configured with the hardware shown in Figure 1. To a programmable laboratory microwave oven (1000 W) equipped with an infrared (IR) thermometer and an external thermometer, non-absorbent microwave material tube coil was placed in front of the infrared thermometer. Polyethylene or silicone tubes may be used. Nylon wire was used to adjust the required shape of the coil in order to fix its correct position in relation to the infrared thermometer. This tube coil constitutes the reactor. This coil is powered by a triple channel peristaltic pump, from a cylindrical vessel, dimensions 60 cm high and 7 cm in diameter, containing soy oil and alcohol/water/KOH mixtures. Three pumping points were provided in the cylindrical vessel, in the middle, at the top and at the bottom (flow rates a, b and c respectively, see Figure 1). A mixture of soy oil (350 ml) and 150 ml of alcohol/water mixtures with the presence of 1% KOH (w/v) is added to the vessel. The diameter of the tubes of the peristaltic pump was chosen so that the flow rates (in mL/min.) respect the ratio of 0.3: 0.4: 0.3 for a, b and c. The need for three liquid collection points and their positioning, with the configuration shown, is due to the known effects of the phase order inversion behavior at an intermediate point of the reaction coordinate, which does not mean the end of the reaction (see Chapter 8). This configuration is also compatible with density values higher or lower than

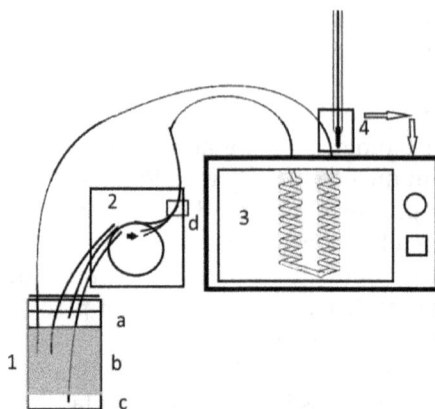

Figure 1. Experimental configuration for the preparation of alkyl esters of fatty acids from edible oils and water/alcohol mixtures, in a continuous reactor with microwave heating. A three-channel peristaltic pump (2) connected to the reagent reservoir (1) with sources at the top (a), center (b) and bottom (c) of the vessel. The flows are mixed in the 3: 1 (d) joint to form an emulsion that is directed to the reactor, a coil of plastic tube inside the microwave oven. The coil is positioned in front of the microwave's infrared thermometer. The temperature at the outlet of the reactor is monitored by a thermocouple and transmitted to the microwave processor. The sink of the reaction product (outlet) is directed to the middle of the reagent vessel. For more details on the experiment, see the text.

that of oil, caused by the presence of water or contaminants in the mixture with alcohol. With this configuration, the triple pumping system allows the complete FAME preparation reaction to be achieved with any phase order configuration. The phase inversion in the reaction coordinate occurs due to the formation of glycerol, a triol. As the reaction proceeds, it starts to mix in the alcoholic phase. The higher density value for glycerol (d = 1,267) causes the alcoholic phase to pass to the bottom of the container at a certain point, unless a very high proportion of alcohol to oil is used. A three-to-one tube adapter converts the triple flow into a single flow that is introduced into the microwave oven coil and subject to microwave irradiation (part 3 in Figure 1). Whatever the flow rate defined in the peristaltic pump, it satisfies the 0.7: 0.3 ratio between organic phase/alcoholic phase, respectively. The initial flow composition will be a mixture of oil and ethanol/water, after the phase inversion, it is oil mixed with FAME and mixture of alcohol, water and glycerol, in the form of an emulsion. A thermocouple-type thermometer was placed at the outlet of the coil, outside the microwave oven to measure the flow temperature at the outlet of the reactor. Since the microwave is programmable, the microwave radiation power has been configured so that the temperature inside the plastic coil reactor does not exceed 100°C and the temperature at the sensor at the output does not exceed 80°C. The oven microprocessor makes the necessary adjustments to the magnetron power, in order to meet these conditions.

Table 1 shows some results obtained in the transesterification of soybean oil, using alkaline catalysis with 1% KOH dissolved in the alcoholic fraction

Table 1. Approximate content of alkyl fatty acid esters (FAME) prepared from soy oil[1] with alcohol/water mixtures, catalyzed with strong base (KOH), under microwave irradiation at a temperature of 80–100°C, in different time periods.

Exp.	Conditions			Time (min)				Notes
	Alcohol	Water/ Alcohol (%)		10	20	30	40	
1	CH_3OH	0	Ester Yield (%)	82	86	88	99	
2	CH_3OH	4		79	84	87	91	Total flow equal to 2 mL/min
3	CH_3OH	8		66	74	82	88	
4	CH_3OH	16		48	52	60	70	
5	CH_3CH_2OH	4		81	95	97	99	

1. Oil acidity less than 0.8%.

in a continuous process with a flow rate of 2 mL/min and a fixed reaction period equal to 1 h. After this period, it was registered if the phase inversion occurred and the FAME yield was determined by HPLC, under conditions described in Chapter 8. Concomitantly analyzes in GC by the method EN 14105 (UE 2003) were also obtained (Varejão and Cruz Costa 2015).

The results in Table 1 show the possibility of efficiently carrying out the transesterification reaction, catalyzed by a strong base, with efficient preparation of the corresponding ester using microwave irradiation heating in a continuous reactor. The results also indicate the possibility of using methanol with high levels of water in the present test indicating that even levels of about 3–4% do not significantly affect the reaction. For low to moderate water content (1–4%), reaction times of 30–70 min at 80–100°C seem sufficient to obtain biodiesel with FAME purity greater than 96%. For higher water levels (experiment 4), longer periods of time seem to be necessary, however, it is not common to find alcohols with these water levels on the market. However, the result suggests that non-rectified alcohols, with high water content, obtained from different areas of human activity may be used in the process. The use of ethanol azeotrope (4% water) does not present any difficulty in the reaction, with the preparation of FAME being obtained in just a 30 min reaction. This result may contribute to a greater use of low-quality alcohol/water mixtures, a contribution to the preparation of more environmentally friendly biodiesel.

5. Use of biodiesel

5.1 Mixture of biodiesel with fossil diesel

In Europe, diesel fuel is sold on the market with a content of 7% FAME, although in the past there was an intention to increase its content to values of 15%, the environmental damage resulting from over-cultivation of oil plants added to that was being caused by the already high demand by the food industry, which led to this intention being postponed. With contents of 7% of incorporation, the objective of reducing emissions of solid particles in the engine is partially achieved, with no major problems associated with its use in most engines and fuel supply systems.

A problem may arise, however, from mixing biodiesel with fossil diesel. The mixing must be done so that the final product is perfectly homogenized. In some cases, small producers and personal users do not have the means necessary for their complete homogenization, and associated problems may arise. Although at first glance the process appears to be free of major problems, it is usually not the case. Especially with the use of large volumes of both components, problems arise if the mixture is not completely homogenized with the use of mechanical stirrers. With smaller volumes, homogenization can be done simply by mixing in clean vessels with rotary shakers. The simple addition of the two components in high volume deposits without any agitation, causes the greater density of FAME (d = 0.880) compared to fossil diesel (d = 0.835), to deposit itself at the bottom of the vase. Especially if the vessel has a high volume, diffusion homogenization can happen very slowly, involving a long period of time. If it is not subject to homogenization, the use of the final fraction may lead to problems in some types of fuel supply systems and engines. It turns out that the unmixed fraction the biodiesel content may reach values of 40% by volume or even higher.

Depending on the age and use of the fossil diesel deposits, either at filling stations or in the vehicles themselves, the interior surface has a thin to thick layer of high molecular weight hydrocarbon deposits, with a waxy appearance, white in color and with low solubility in fossil diesel, consisting mainly of a complex mixture of hydrocarbons in C8–C14. In Figure 2, a section of a relatively new diesel tank is shown, showing the growing waxy material layer. For older vessels, it is common to see formations with a similar aspect to stalagmite/stalactite but of a hydrocarbon nature, and of much larger dimensions.

The solubility properties of biodiesel, an ester in its chemical nature, are far superior to the hydrocarbons that make up fossil diesel. When a biodiesel /diesel mixture is introduced into one of these deposits, especially if it

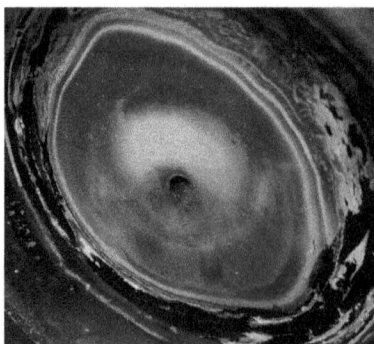

Figure 2. Section of a fuel tank of an automobile equipped with a diesel engine showing the presence of white waxy deposits on its surface, formed by the precipitation of hydrocarbons of high molar mass of low solubility in diesel.

contains a high biodiesel content and has low homogeneity, the waxy coatings are solubilized and then deposited, resulting in wax emulsions in biodiesel/diesel mixtures with a similar appearance to milk. These mixtures clog the fuel filters, and if they reach the high-pressure fuel pumps they usually lead to their destruction resulting in need of replacement, an arrangement that is often expensive. Other components such as injectors may be affected, as it is sometimes necessary to clean the entire fuel line.

In order to avoid the problems described above, complete homogenization of biodiesel/diesel mixtures must be ensured, and the introduction of the use of biodiesel in any vehicle, especially in older vehicles, be carried out in a phased manner with increasing concentrations of biodiesel. From experience of use in a wide variety of heavy agricultural equipment and in commercial vehicles, the use of 10% biodiesel/diesel mixtures, which are completely homogenized, do not present problems in most vehicles. Under these conditions, the wax coatings are slowly solubilized, allowing over time for the increase of the biodiesel content in the mixture, limited, however, to the design of the injection system.

For small biodiesel producers that do not have adequate mixing units, addition techniques have been developed that allow homogenization in the fuel tank itself. For this purpose, some precautions must be taken to avoid problems. The method of adding increasing quantities mentioned above must be respected. An example that has been shown to be effective is to add pure biodiesel to the tank in question, ensuring that the fuel gauge is working correctly. This way, knowing the volume of the tank, its reading allows to evaluate the total volume of fuel in the vehicle at a given moment. Biodiesel must always be added in the last place, and the movement of the vehicle ensures the homogenization of the mixture. The initial gradual increase in the concentration of biodiesel should be done, sequentially, especially for older vehicles or with a high number of kilometers driven—at the beginning a content of 5%, followed by 10%, 15–20%, to 25% and up to 30%. If this sequence is respected, at least 2–3 deposits for each concentration value, the waxes will dissolve gradually. Biodiesel is generally marginally denser than fossil diesel and tends to move to the bottom of the deposits, however the density difference is low, which means that with normal movement of the mixture, homogenization of the mixture can be achieved in a short period of time. After the adaptive sequence is made, generally the continuous use of 15% biodiesel mixtures used does not give rise to any problem. The use of levels above 15%, up to 30% is possible, but it has to be evaluated in relation to the type of pump and injectors, especially by the presence of rubber components usually rings or others. Biodiesel, due to the differences in solubility mentioned above, can lead to the swelling of some rubber rings leading to the eventual failure of high pressure pumps and even other components in the fuel line, such as low pressure and high pressure pumps and injectors. One possibility is to switch from rubber to materials compatible with biodiesel. Careful knowledge of the materials used in the fuel and injection line is necessary and it is recommended to obtain information from the manufacturers. In specific engines, which do not have any rubber or polymer component, the entire tank system, fuel line and injector being solely metallic, it is possible to use 100% biodiesel as a single fuel without any noticeable problems during years of work. For the others, the recommendation is not to exceed a maximum biodiesel content of more than 30% (v/v).

5.2 Use of biodiesel in heating units

Biodiesel can be mixed with heating oil or other fuel to be burned in heating units for domestic and industrial installations. For use of heating oil/biodiesel mixtures up to about 30%, direct use generally does not bring any inconvenience, thus it is possible to use conventional burners.

Conventional burners are unable to ignite flame with heating oil/biodiesel mixtures at concentrations above about 30%. In order to use higher concentrations, boilers generally require modifications to the initiation hardware, usually starting with a lighter fuel such as gasoline, and after some time the change to biodiesel is carried out automatically by the system. Biodiesel can eventually be heated by the system itself to improve its burning properties. In this case changing the burner is necessary.

5.3 Biodiesel stability

Biodiesel or FAME is a mixture of fatty acid esters prepared from lipid residues, originating predominantly as a residue from the preparation of human foods, mainly in frying food. Restaurants and hotels are major producers of the lipid residue that is collected for processing. The oxidative stability of FAME is parallel to that of edible oils. The most common types of edible oil found in residual oils reflect the composition of the oils originally chosen in the preparation of commercial mixtures. For reasons of price, market transactions and availability, food oils for frying around the world are mainly mixtures of genetically modified soy oil mixed with palm oil. These can reach the market at very competitive prices, often lower than vehicle fuels, at which high rates are imposed in their marketing. Local regions can use different oil compositions, originated by local availability, such as the use of rapeseed and sunflower oils, in the Iberian Peninsula the use of olive oil is widely accepted, however globally the first mixtures are predominant.

The problem with soybean oil is that it is a lipid that has a relatively high content of polyunsaturated fatty acids, such as linoleic and linolenic acids, which are healthy if eaten without heating, but are not very stable to heating, especially in the presence of oxygen. The frying process of food is generally carried out at temperatures of 180 to 200°C, conditions that lead to rapid oxidative degradation of the same, which leads to rancification and/or polymerization to resins of high molar mass. To prevent or at least slow down the rate of degradation, antioxidant substances are added, such as di-t-butylmethylphenol (BHT) (Esposo et al. 2015). In any case, they will degrade and quickly lead to the formation of triacylglycerol decomposition products (Liu et al. 2019). The initial steps of the degradation process of triacylglycerols are shown in Figure 3.

Bis-allylic positions are specially activated by resonance with the C= C double bonds and exist in polyunsaturated fatty acids such as linoleic and linolenic. Hydroperoxides are formed in the presence of oxygen at elevated temperature (Hwang et al. 2013). In order of reactivity the most reactive positions are the double activated allyl and then the simple allyl positions. Saturated fatty acids are quite resistant to thermal oxidative degradation. Peroxides are the initial products of oxidation and are very unstable substances that react quickly intramolecularly with existing double bonds, leading to cuts in the bonds of the fatty acid bonds to produce aldehydes. These are easily converted into short-chain carboxylic acids, making the so-called rancification of fats (Thomasson et al. 2020). The analytical parameter content of total polar compounds is high at this point and this parameter includes low and high molecular weight carboxylic acids (Baena and Calderon 2020).

The other route of degradation of the triacylglycerols leads to the formation of oligomers and in the final stages and even polymers. Hydroperoxides easily

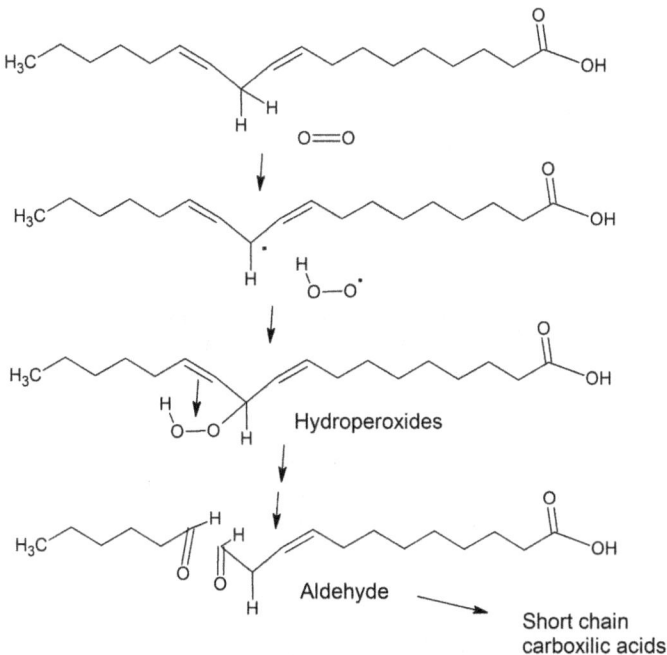

Figure 3. Initial stages of oxidative degradation of triacylglycerols. Fatty acids that have bis-allyl positions are especially susceptible to rapid oxidation by atmospheric oxygen to produce hydroperoxides. These are unstable substances that react with the close carbon-carbon double bonds to produce oxidized intermediates whose decomposition mainly originates aldehydes. Aldehydes are easy to oxidize by molecular oxygen, and converted to peroxyacids that decompose into low molecular weight carboxylic acids.

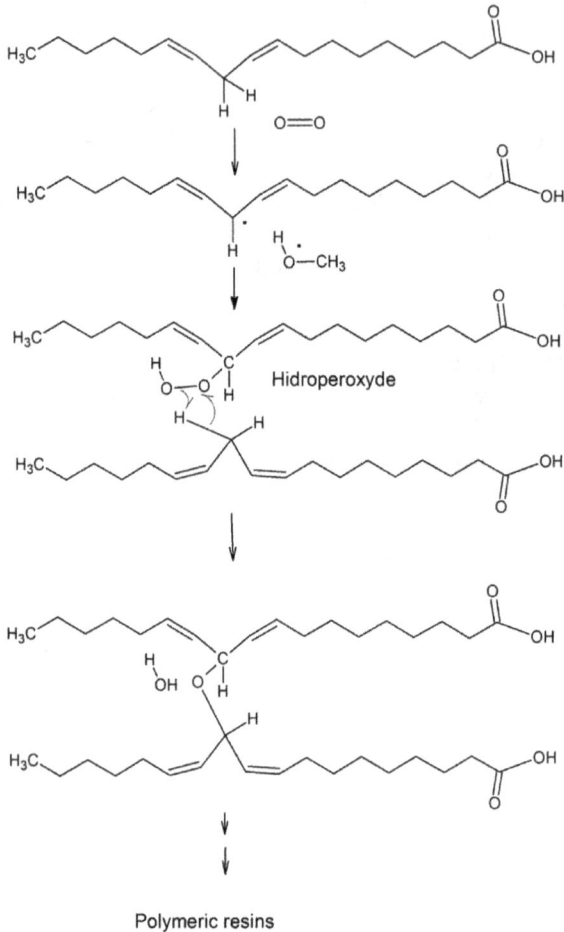

Figure 4. Degradation of fatty acid hydroperoxide with double carbon bonding via intermolecular attack to form dimeric species with high molar mass. The oligomerization and eventual subsequent polymerization occur through the formation of C-O-C ether bonds, together with C-C bonds.

generate radical intermediates that can react intermolecularly leading to the formation of substances of high molar mass, as shown in Figure 4. In this case the result is the formation of triacylglyceride oligomers, of high molecular weight, with a considerable increase in oil viscosity (Kmiecik et al. 2020). The concentration of oligomers begins to increase in aerobic conditions, catalyzed by light and high temperatures.

The freshly prepared biodiesel undergoes the oxidative degradation mentioned above and which, without prior warning, can degrade the product to values off its normal and legal parameters with probable problems with vehicles that are mainly centered on fuel filters, injectors and high pressure pumps.

The generation of organic acids increases with the acidity value of biodiesel and can give it, over the period of time from production to consumption, corrosive properties, even if initially its acidity characteristics were within the legislated parameters.

For these reasons, the biodiesel produced must be stored under anaerobic conditions, such as a carbon dioxide atmosphere, in the absence of light and at the lowest possible temperature. The addition of synthetic antioxidants can be useful and change the composition of fatty acids by mixing lipids rich in saturated fatty acids, more resistant to oxidation, if available, may be techniques that help to prevent the occurrence of these problems (Jose and Anand 2016).

6. References

Abas, N. A., Yusoff, R., Aroua, M. K., Aziz, H. A. and Z. Idris. 2020. Production of palm-based glycol ester over solid acid catalysed esterification of lauric acid via microwave heating. Chemical Engineering Journal 382: 122975.

Ali, M. A. M., Gimbun, J., Lau, K. L., Cheng, C. K., Vo, D. V. N., Lam, S. S. and R. M. Yunus. 2020. Biodiesel synthesized from waste cooking oil in a continuous microwave assisted reactor reduced PM and NOx emissions. Environmental Research 185: 109452.

Baena, L. M. and J. A. Calderon. 2020. Effects of palm biodiesel and blends of biodiesel with organic acids on metals. Heliyon 6: e03735.

Barbosa, S. L., Ottone, M., Freitas, M. D., Lima, C. D., Nelson, D. L., Clososki, G. C., Caires, F. J., Klein, S. I. and G. R. Hurtado. 2019. Synthesis of phenyl esters using SiO2-SO3H catalyst in conventional heating and microwave-irradiated esterification processes. Journal of Nanoscience and Nanotechnology 19: 3663–3668.

Binnal, P., Amruth, A., Basawaraj, M. P., Chethan, T. S., Murthy, K. R. S. and S. Rajashekhara. 2020. Microwave-assisted esterification and transesterification of dairy scum oil for biodiesel production: Kinetics and optimisation studies. Indian Chemical Engineer. In press.

Bren, M., Janezic, D. and U. Bren. 2010. Microwave catalysis revisited: An analytical solution. Journal of Physical Chemistry A 114: 4197–4202.

Davies, E., Deutz, P. and S. H. Zein. 2020. Single-step extraction-esterification process to produce biodiesel from palm oil mill effluent (POME) using microwave heating: A circular economy approach to making use of a difficult waste product. Biomass Conversion and Biorefinery. In press.

de Oliveira, A. D., de Oliveira, D. T., Angelica, R. S., Andrade, E. H. D., da Silva, J. K. D., Filho, G. N. D., Coral, N., Pires, L. H. D., Luque, R. and L. A. S. do Nascimento. 2020. Efficient esterification of eugenol using a microwave-activated waste kaolin. Reaction Kinetics Mechanisms and Catalysis 130: 633–653.

EN 14105. Fat and oil derivatives—Fatty Acid Methyl Esters (FAME) determination of free and total glycerol and mono-, di-, tri-glyceride content. European Committee for Standardization: Management Centre, rue de Stassart 36, B-1050 Brussels, 2003.

Esposto, S., Taticchi, A., Di Maio, I., Urbani, S., Veneziani, G., Selvaggini, R., Sordini, B. and M. Servili. 2015. Effect of an olive phenolic extract on the quality of vegetable oils during frying. Food Chemistry 176: 184–192.

Gihaz, S., Kanteev, M., Pazy, Y. and A. Fishman. 2018. Filling the void: Introducing aromatic interactions into solvent tunnels to enhance lipase stability in methanol. Applied and Environmental Microbiology 84: e02143–18.

Hwang, H. S., Winkler-Moser, J. K., Bakota, E. L., Berhow, M. A. and S. X. Liu. 2013. Antioxidant activity of sesamol in soybean oil under frying conditions. Journal of the American Oil Chemists Society 90: 659–666.

Imanparast, S., Faramarzi, M. A. and J. Hamedi. 2020. Production of a cyanobacterium-based biodiesel by the heterogeneous biocatalyst of SBA-15@oleate@lipase. Fuel 279: 118580.

Jose, T. K. and K. Anand. 2016. Effects of biodiesel composition on its long term storage stability. Fuel 177: 190–196.

Keglevich, G., Kiss, N. Z., Mucsi, Z. and T. Kortvelyesi. 2012. Insights into a surprising reaction: The microwave-assisted direct esterification of phosphinic acids. Organic & Biomolecular Chemistry 10: 2011–2018.

Khedri, B., Mostafaei, M. and S. M. S. Ardebili. 2019. A review on microwave-assisted biodiesel production. Energy Sources Part A-Recovery Utilization and Environmental Effects 41: 2377–2395.

Kmiecik, D., Fedko, M., Siger, A. and B. Kulczynski. 2020. Degradation of tocopherol molecules and its impact on the polymerization of triacylglycerols during heat treatment of oil. Molecules 24: 4555.

Lee, S. B., Jang, H. S. and B. H. Yoo. 2018. Preparation of waste cooking oil-based biodiesel using microwave energy: Optimization by box-behnken design model. Applied Chemistry for Engineering 29: 746–752.

Liu, X. F., Wang, S., Masui, E., Tamogami, S., Chen, J. Y. and H. Zhang. 2019. Analysis of the dynamic decomposition of unsaturated fatty acids and tocopherols in commercial oils during deep frying. Analytical Letters 52: 1991–2005.

Mazo, P. C. and L. A. Rios. 2010. Esterification and transesterification assisted by microwaves of crude palm oil homogeneous catalysis. Latin American Applied Research 40: 337–342.

Muanruksa, P., Dujjanutat, P. and P. Kaewkannetra. 2020. Entrapping immobilisation of lipase on biocomposite hydrogels toward for biodiesel production from waste frying acid oil. Catalysts 10: 834.

Mujtaba, M. A., Masjuki, H. H., Kalam, M. A., Noor, F., Farooq, M., Ong, H. C., Gul, M., Soudagar, M. E. M., Bashir, S. and I. M. R. Fattah. 2020. Effect of additivized biodiesel blends on diesel engine performance, emission, tribological characteristics, and lubricant tribology. Energies 13: 3375.

Multari, S., Marsol-Vall, A., Heponiemi, P., Suomela, J. P. and B. R. Yang. 2019. Changes in the volatile profile, fatty acid composition and other markers of lipid oxidation of six different vegetable oils during short-term deep-frying. Food Research International 122: 318–329.

Nayak, S. N., Bhasin, C. P. and M. G. Nayak. 2019. A review on microwave-assisted transesterification processes using various catalytic and non-catalytic systems. Renewable Energy 143: 1366–1387.

Nguyen, H. C., Wang, F. M., Dinh, K. K., Pham, T. T., Juan, O. R. Y., Nguyen, N. P., Ong, H. C. and C. H. Su. 2020. Microwave-assisted noncatalytic esterification of fatty acid for biodiesel production: A kinetic study. Energies 13: 2167.

Patel, A., Arora, N., Pruthi, V. and P. A. Pruthi. 2019. A novel rapid ultrasonication-microwave treatment for total lipid extraction from wet oleaginous yeast biomass for sustainable biodiesel production. Ultrasonics Sonochemistry 51: 504–516.

Sharma, R. K., O'Neill, C. A., Ramos, H. A. R., Thapa, B., Barcelo-Bovea, V. C., Gaur, K. and K. Griebenow. 2019. Candida rugosa lipase nanoparticles as robust catalyst for biodiesel production in organic solvents. Biofuel Research Journal-BRJ 6: 1025–1037.

Simsek, S. and S. Uslu. 2020. Comparative evaluation of the influence of waste vegetable oil and waste animal oil-based biodiesel on diesel engine performance and emissions. Fuel 280: 118613.

Thomasson, M. J., Diego-Taboada, A., Barrier, S., Martin-Guyout, J., Amedjou, E., Atkin, S. L., Queneau, Y., Boa, A. N. and G. Mackenzie. 2020. Sporopollenin exine capsules (SpECs) derived from Lycopodium clavatum provide practical antioxidant properties by retarding rancidification of an omega-3 oil. Industrial Crops and Products 154: 112714.

Varejao, J. M. T. B. and M. C. Cruz Costa. 2015. New on-column GC sample injection techniques: Making holes in separation capillaries. Scientia Chromatographica 6: 277–285.

Wijaya, H., Sasaki, K., Kahar, P., Quayson, E., Rachmadona, N., Amoah, J., Hama, S., Ogino, C. and A. Kondo. 2020. Concentration of lipase from Aspergillus oryzae expressing Fusarium heterosporum by nanofiltration to enhance transesterification. Processes 8: 450.

Wong, K. Y., Jo-Han, N., Chong, C. T., Lam, S. S. and W. T. Chong. 2019. Biodiesel process intensification through catalytic enhancement and emerging reactor designs: A critical review. Renewable & Sustainable Energy Reviews 116: 109399.

Index

For Product Safety Concerns and Information please contact our EU
representative GPSR@taylorandfrancis.com
Taylor & Francis Verlag GmbH, Kaufingerstraße 24, 80331 München, Germany

www.ingramcontent.com/pod-product-compliance
Lightning Source LLC
Chambersburg PA
CBHW060556220326
41598CB00024B/3116